ARITHMETIC

ARITHMETIC

Paul Lockhart

The Belknap Press of
Harvard University Press
Cambridge, Massachusetts
London, England
2017

Library of Congress Cataloging-in-Publication Data

Names: Lockhart, Paul, author.
Title: Arithmetic / Paul Lockhart.
Description: Cambridge, Massachusetts : Harvard University Press, 2017. |
    Includes index.
Identifiers: LCCN 2016055136 | ISBN 9780674972230 (alk. paper)
Subjects: LCSH: Arithmetic. | Arithmetic--History.
Classification: LCC QA115 .L713 2017 | DDC 513--dc23 LC record available
at https://lccn.loc.gov/2016055136

CONTENTS

Dear Reader   vii

Things   1
Language   5
Repetition   10
Three Tribes   15
Egypt   25
Rome   32
China and Japan   41
India   48
Europe   75
Multiplication   87
Division   117
Machines   136
Fractions   151
Negative Numbers   180
The Art of Counting   197

Afterword   215
Index   217

Dear Reader,

It's fun to count and arrange things. We like doing it, and we have even developed it into something of a folk art. This art is called *arithmetic*. Arithmetic is the skillful arrangement of numerical information for ease of communication and comparison. It is a fun and enjoyable activity of the mind and a relaxing and amusing pastime—a kind of "symbol knitting," if you will. But please understand that's all it is. Being good at it doesn't make you particularly smart or mathematically inclined or anything like that. Similarly, being unskilled at arithmetic does not mean that you are stupid or that you do not have a mathematical mind. Arithmetic is just like any other craft; you can get good at it if you want to, but it is no big deal either way. My hope is that by reading this book you will be inspired to try it out for yourself and to experience firsthand the simple pleasure and satisfaction that comes with numerical fluency. Have fun!

THINGS

We find ourselves in a world full of things: plants and animals, rocks and trees, the stars in the sky and the sand on the beach. We are surrounded by multitudes. And we *count* them. Why exactly do we bother counting things? The truth is, we usually don't. Most situations do not really call for any careful counting, just a sense that "we have plenty enough" or "we're way short." The main reason why we count is to *compare*. Usually these comparisons are easy to make: "So these together are about sixteen bucks . . . Oh, I have a twenty. No problem." In fact, most transactions these days sound less like "sixteen seventy-one out of twenty, that's three twenty-nine. Have a nice day," and more like "Beep. Boop. *Swipe*." Most of the time we don't need to do much actual counting. Still, there are times when we do want an accurate idea of exactly how many what-have-yous we happen to have. What sorts of things do we keep track of, and why?

One of the oldest is *time*. We can imagine meetings between prehistoric tribes being scheduled for so-many moons hence, or scratches carved in a tree counting the days the hunting party has been away. We certainly do keep track of time—too much, if you ask me. Money, of course, is another one. People definitely keep careful track of that stuff (don't get me started). Property also: "Are all of my marbles here in my marble bag?" "I hope we have enough silverware." Land, grain, cattle, the inventories of the ages. Seven years of famine, seven of plenty. Boy, do we like to count!

Then there is curiosity counting—counting for the sake of pure wonder. "How many stars are in the Milky Way?" "For how many seconds can I hold my breath?" Or the more mathematical: "How many ways can I arrange the books on my bookshelf?"

Whatever the purpose or rationale, we occasionally find ourselves wanting to know how many of something there are. This is where arithmetic begins, with *desire*. There's no point counting something you don't care about. Don't ever do that. It's boring, and it will make you hate counting. There are actually a lot of people who hate arithmetic (far too many to count!), and it makes me sad. Usually it's because they were made to do something they weren't interested in doing. Let's not have that be you.

The point is that there are things, and we sometimes want to count them. Actually, there are some quantities that we don't so much count as *measure*. What we do when we measure something—and this could be an amount of milk, or someone's height, or a piece of land—is we take some agreed-upon amount, known as a *unit* (e.g., a cup, an inch, or an acre), and we count the number of those. This has the effect of turning something smooth and continuous into something discrete and lumpy. So let's confine ourselves to thinking about counting distinct individual items.

As I said before, counting is really all about comparison. Even when you count something out of curiosity—say you happen to have thirty-two pennies—the only sense in which you "know" how many you have is by comparison with certain other amounts (e.g., thirty), which have special names in your language and serve as familiar benchmarks of size. Just saying the word *thirty-two* means you have unconsciously performed a comparison: you have a little more than three double handfuls.

Comparison is always the issue. "Do we have enough?" "Is that all going to fit?" "There were more of them before. Where'd they all go?" We count so that we can compare amounts. Now, it would be great if we could just look at two collections and know which one was bigger, as though we had a "number sense" akin to smell or taste. Maybe there are creatures like that who can just *tell*. Occasionally, some humans come along (usually severely neurologically impaired ones) who can do essentially that, but most of us are pretty bad at it. I think pretty much everyone can perceive "three-ness"—knowing that there are three things without having to literally count them—but past about six or seven it starts to get a little dicey.

Try out your ability to directly perceive quantity.
I wonder if we can get better at it with practice?

One problem is that we are a little too good at pattern recognition. Our brains are quite good at storing patterned information; we don't do so well with random spatters. Here are two examples of sixness:

I can't really say that I've got sixness perception. I had to count the first collection. But the second one was immediate. We've all seen that arrangement a million times. Dice players don't need to count the pips. We know those shapes like the backs of our hands. Better, in fact.

So it helps if the things we want to count are somewhat organized. Nobody wants to count a bunch of random objects strewn all around the room. For one thing, the act of counting strongly relies on memory and time sense: "Which ones have I already counted?" Everyone who has ever lived knows the sickening feeling of *losing count*—forgetting where you were and having to start over. This may in fact be the single most common shared experience of humanity. We are all, under the skin, people who have at one time or another lost count. This is especially easy to do if you can't remember which things you have already counted because they all look the same and are scattered about topsy-turvy and higgledy-piggledy.

Not that we don't lose count anyway, even when things are nicely organized. The truth is that we're just not very good at boring, repetitive activities. Such things simply don't hold our attention, and our minds tend to wander. We're just too darn smart and interesting, I guess.

OK, so we don't have instant number perception past five or so, and we have trouble with chaos and disorder. Also, we are careless daydreamers who can't be trusted to perform one simple mind-numbing and repetitive task. So what do we do? We come up with art and science, that's what. We develop a creative and entertaining craft of counting well, using our pattern-seeking minds.

How would you arrange a number of rocks
to make it easy to tell if it is even or odd?

LANGUAGE

One way to think about counting is that we are measuring the size of a collection. Implicit in this is the recognition of the fact that the number of things is independent of how they are arranged. If I put seven marbles in my pocket, it doesn't matter how I jiggle them around; when I take them out, there will still be seven (assuming I don't jiggle them so hard they shatter, and that I don't have a hole in my pocket). At some point in our mental development, we become aware of this *number permanence*. What this means is that we get to arrange and rearrange our collections in various ways without disturbing the information we are interested in—namely, how many things the collection has in it.

This is all very well and good when the things are relatively easy to hold, carry, and manipulate (e.g., marbles, coins, or jelly beans). Often, however, the things we want to count are inconveniently large, far away, or, even worse, moving. And sometimes, as in the case of time, what we wish to count is ephemeral and fleeting.

What we do in these cases, of course, is to substitute. We replace the things we are actually interested in counting by something that is easier to work with. We make a *representation* of the situation and work with that instead.

It is hard to say when this idea first originated, but it is clear from cave paintings and carved figurines that we have been representing things by other things for quite some time.

Although there are instances of this behavior in other species (primate gestures, bumblebee dances, etc.), nothing comes close to the extent and frequency with which humans perform these substitutions, often unconsciously. Language itself is nothing other than a vast interconnected set of sense-memory substitutions, with words used to represent instances of similar sensory experiences. It is safe to say that representation is one of the fundamental ingredients of consciousness.

In fact, we are so good at making representations that we often lose track of the distinction between the representation and the thing itself. The word *cat*, for instance, is a string of three letters. It does not have a tail and it does not purr. The string of letters is a *code*—it stands for the thing with a tail that purrs. This is what Juliet is referring to when she muses to herself,

> What's in a name? That which we call a rose
> By any other name would smell as sweet.

In particular, when we start designing arithmetic systems, it will be crucial to keep clear the distinction between a number and its representation. A lot of people are walking around in a muddle about this, and I don't want you to be one of them.

Anyway, at some point—let's say two hundred thousand years ago—people started representing things by other things. One big advantage of doing this is *portability*. If I use pebbles to represent my sheep, I can then easily carry them around in a bag and use them to buy and sell and keep track of my flock, without having to carry around any actual sheep. (This is not only much easier on my back but is also vastly preferred by the sheep.) This is undoubtedly the way money originated. Similarly, a football coach can use a diagram of Xs and Os to represent players and can then move the symbols around on paper instead of having to position actual large and cumbersome football players.

A caveman wanting to convey some vital numerical information, such as how many caribou he spotted down by the water hole or (perhaps more important) how many lions, could quickly gather some rocks or berries, matching each animal with its more portable representative, and return with the information literally in the palm of his hand. That's what numbers are, really: *information*. And the point is that the replacement of caribou by rocks doesn't change the information. It does, however, change the ease with which the information is held and transmitted. So that's why we do it. The history of language (and arithmetic in particular) is a history of increasingly abstract and portable substitutions: from actual antelope to drawings of antelope, to rocks, sticks, or fingers, to scratches on bones, to verbal utterances (i.e., words), to abstract symbols.

"three" 3

My guess is that each of these developments in the evolution of language arose from the need to explain and communicate information. Mathematics, in particular, is an explanatory art form, and ultimately all of its structures arise as "information carriers" for the purpose of explaining a pattern or idea. Numbers are simply the most basic kind of mathematical information.

So in this book I'm going to tell you the story of arithmetic, and in a large part this is a history of representation—different choices and strategies for organizing and communicating numerical information. Throughout this story, it will be important to keep in mind the distinction between a number in and of itself, as opposed to the way in which it is being represented. For concrete things like cats this isn't usually too difficult—I don't think too many people are going around petting the letters C, A, T. But numbers are a bit more elusive. I can easily understand how a conflation of the number three and the symbol 3 can occur. If 3 is merely a symbol representing the number three, then what exactly is three itself, independent of its representation? I've held cats and petted them, but how can I feel or touch *three*?

Of course, it's not as bad as some abstract ideas, such as hope or love—at least I can hold three oranges or three feathers in my hand. Nevertheless, *threeness*, that property of collections, is a bit intangible and wispy. Be that as it may, three has properties and truths (such as being odd) that are independent of its representation.

It is especially vital that we be clear about this distinction if we are going to examine representation systems other than those with which we are familiar. For example, the number thirteen does not "start with a one and end with a three." Numbers don't start or end at all; they just *are*. The Hindu-Arabic decimal place-value representation is the thing that starts and ends and has a look to it. Thirteen does a lot of things—it is prime, it is a sum of two square numbers (namely four and nine), and it is odd—but it doesn't "look like" anything. I myself like to think of numbers as creatures of some kind that exhibit various behaviors, and this helps me play and work with them, but I don't want to get stuck into thinking of them as having any particular visual form— or rather, I understand that the form they take is up to me, and I can represent them well or badly depending on my choices.

Among my favorite ways to represent quantities is to imagine them as piles of rocks. As we've seen, depending on the size of the collection there will be a myriad of possible visual arrangements, some more organized (and thus easier to remember and recognize) and some less so.

> *For each number of rocks up to twelve,*
> *what is your favorite arrangement?*

For example, the arrangements of pips on a die are particularly simple and easy to remember:

(By the way, the five shape ⚄ is called a *quincunx*, a word one does not often get the opportunity to use. Naturally, I jumped at the chance.)

If the only numbers we ever wanted to communicate were the numbers one through six—as in a game of Chutes and Ladders, for instance—then these dice arrangements would constitute a perfectly valid and convenient number language. It's the same with fingers. When someone (such as a referee) holds up two fingers of one hand and all five fingers of the other, everyone can immediately recognize the finger pattern of seven. The real problem in designing an intelligent numerical representation system is what to do when the quantities get big. And as we will see, the problem is not just that we run out of fingers. The problem is that we run out of *memory*.

*Can you invent a way to represent the numbers
one through twenty using only the fingers of one hand?*

REPETITION

Undoubtedly the oldest and simplest way to represent numbers is by using tally marks—an organized series of strokes or scratches, one for each item to be counted. The number four, for instance, would be represented as IIII. One of the earliest known examples of this is the Ishango bone, dated from about 18,000 BC.

Though it is unclear what was being counted, it certainly appears to be a record of *something*. (Of course, it's possible that these scratches are just a record of idle boredom or mere decoration.) In any case, people have been representing numbers in this way from time immemorial.

The tally mark scheme has a number of things going for it. First of all, it is organized. We replace the possibly random assembly of objects to be counted with a nice, regular sequence of more or less identical, evenly spaced strokes. What could be simpler or more natural? This is the traditional way of keeping track of the score in many games, as well as the number of days one has been in prison. Tally marks are simple, portable, and convenient.

As always, the trouble starts when the numbers get big. For small quantities (say five or fewer), we can easily distinguish these representations and immediately perceive and compare amounts. Nobody needs to count the individual strokes in the tally mark representation of four; we can just tell there are four of them. But as we've seen, we tend to lose this ability for larger quantities. Even though the tally mark system is simple and easy to use, it still doesn't solve the "perception problem."

Which is larger, IIIIIIIIII or IIIIIIIII?

There are many ways we can improve this system, the most familiar being the so-called five-barred gate. In this scheme (which is also older than recorded history), we make a diagonal slash at every fifth item. For example, the five-barred gate representation of thirteen would be

𝍸𝍸𝍷𝍷𝍷

The diagonal strokes help to break up the monotony and allow us to easily distinguish one amount from another. The tally marks are bundled into little groups of five, each resembling the typical pasture gate of five bars. We can immediately perceive two "handfuls" and three "leftovers." It is now a simple matter to compare two quantities expressed in this way. If one of them has more gates (or handfuls) then it is larger. If they have the same number of gates, then we compare the leftovers. So that is a simple and elegant solution, and most people find it easy to learn and use.

Which is larger, 𝍸𝍸𝍸𝍷𝍷𝍷 *or* 𝍸𝍸𝍸𝍷𝍷 *?*

This is an example of a very general and important technique in arithmetic known as *grouping*. Since long sequences of repeated symbols are difficult to perceive, remember, and compare, we break the list up into little clumps of a fixed size and indicate that in some way. This complicates the language a bit (in that we need a new mark or word or something), but it is almost always worth it. The same idea appears in music, where long sequences of beats are grouped into rhythmic patterns. It's something we seem to do naturally.

One of the principal benefits of this grouping idea is that it saves time. Not only does it make perception and comparison easier, but it also allows us to communicate numerical information more rapidly and accurately.

Once you have decided to put things into groups, the first question is, what size groups? A composer, for instance, needs to decide how to break the rhythmic pulses into measures, among the many options being the waltz pattern of three beats per measure: TA-ta-ta / TA-ta-ta. Other standard choices include two-, four-, six-, eight-, and twelve-beat patterns.

Similarly, an arithmetician will make a choice of grouping size convenient for the counting problem at hand. Entire number languages—words, symbols, and calculating devices— are designed around such choices and become part of the tribal culture. The five-barred gate system, for instance, places emphasis on the number five and makes that particular quantity feel more "solid" or "clean" than other amounts. Nobody really likes leftovers; they feel like unpleasant remnants, or at any rate, details that will have to be remembered and kept track of. How nice to be able to say "four handfuls" instead of "three handfuls and (grumble) two leftovers."

Of course, one's grouping size is a completely arbitrary and personal choice. There is nothing particularly special about the number five, other than the fact that we happen to have that many fingers on each hand. Over the centuries people have experimented with many different grouping sizes in many different contexts. We can still see traces of these choices in the number words of various languages. The words *dozen* in English and *zwolf* in German reveal a history of grouping by twelves (which we still do with inches and eggs), and an ancient choice of twenty as a grouping size still resonates in the words *vingt* (French) and *score* (English).

Aside from the fact that we are curious and easily bored (which has indirectly led to art, science, and technology), another great hallmark of humanity is that we are *lazy*. We get tired of making busloads of tally marks, and we want to find a way out of doing it. The solution is *abbreviation*. This is the way we always use language. Any time we find ourselves saying the same thing over and over, we simply come up with a new, shorter word for it. This is true for both written and spoken language.

Imagine we are shepherds in nearby pastures (say in ancient Greece), and I want to tell you how many of my sheep were eaten by wolves during the night. Let's say we have worked out a system (a language, really) whereby I bang two sticks together for each lost sheep. Just as we tire of looking at endless rows of tally marks, we also aren't very good at keeping track of long sequences of identical sounds. So one idea would be to use a different sound—maybe a somewhat lower pitch—to indicate a group of five. Just as we write ⊁⊁⊁⊁⊁⊁⊁||, we could tap out *clop-clop-clop-click-click* and save time and energy that way.

Similarly, if we get tired of making five-barred gates all day long, we could invent a new shorthand symbol, such as a horizontal stroke –, in place of the more time-consuming ⊁. Then our number above could be written simply as – – –||, or even ≡||. That's the kind of thing we humans do. We get lazy and bored and then we use our intelligence and creativity to come up with clever new ways to get out of doing work. We are all Tom Sawyers on some level. Necessity may be the mother of invention, but boredom is surely the father.

However we wish to communicate numbers—by words, gestures, or in written form—repetition, grouping, and abbreviation are the natural linguistic means. It's up to the users of such languages to decide on a convenient grouping size and how they wish to indicate both groups and leftovers.

So how do we decide on a good grouping size? What makes one size better than another? Of course it depends on what you are counting, who you will be communicating with, and, most important, how large the numbers are going to get.

If you are only counting a small handful of things (e.g., your marble collection), then it doesn't really matter whether or not you organize it into groups or what size groups you use. On the other hand, if the numbers you are using are very large, you may want to put a little thought into choosing an appropriate grouping size.

If your choice of grouping size is too small, say two or three, then it kind of defeats the whole purpose of grouping. At some point, even with the five-barred gate, the numbers will get so large that you will need a ton of groups. So many, in fact, that you will no longer be able to grasp at a glance how many groups you have. You will experience a "higher level" perception problem. As an illustration, can you tell which collection is larger?

JHT JHT JHT JHT JHT JHT JHT JHT   JHT JHT JHT JHT JHT JHT JHT
JHT JHT JHT JHT JHT JHT JHT|| JHT JHT JHT JHT JHT JHT JHT|||

Clearly the thing to do is to organize the gates themselves into groups. And here we have an interesting choice. Each gate represents a group of five individuals; that is, we chose five as our grouping size. But now it is the gates themselves that become the individuals, and for the same reasons as before (perception and comparison), we want to organize *them* into groups. How should we bundle them? Should we group them by fives again or by some different number?

Suppose we decide to group the gates into fours. (This would be natural if we were thinking of the gates as handfuls, so that all our fingers and toes together make four hands' worth.) A nice abbreviation for four gates might be a square shape, □ (the four corners remind me that it stands for four gates). This allows us to write rather large numbers quite easily: □□□ JHT JHT||.

The point is that with any repetition system, as the numbers get larger we will continually need new words and symbols. Each time we hit the next "perception wall" where we start losing track, we will need to decide on a grouping size and a representation for it. Throughout the centuries there have been hundreds of ingenious systems for doing this, some very simple and elegant, others somewhat awkward and annoying. I will be showing you some of my favorites to give you an idea of the range of possibilities.

Can you design your own system of
repetition and grouping?

THREE TRIBES

I want to imagine a period before recorded history—let's say thirty thousand years ago, by the banks of the Nile. (This way I can make up stuff and not worry about being contradicted by actual facts). Let's imagine three groups of early humans living and interacting with each other, each tribe using its own unique number language with its own particular choice of grouping size.

We'll start with the Hand People. In this tribe, the handful—five fingers' worth—is the conventional grouping size, just as in the five-barred gate system. Being a preverbal society, the Hand People communicate using a system of hand gestures. The number one is represented by a single clap, two by two claps, and so on. Five claps are considered difficult to distinguish from four, so a group of five is instead represented by a thump on the chest with a closed fist. Thus, the number that we call seven would be communicated by the gesture: *thump-clap-clap*. The Hand People possess no spoken or written forms of communication.

> *Can you count to twenty in the language of the Hand People?*

Our second tribe is the Banana People. This group expresses its numbers verbally. The word for one is *na*, two is *na-na*, and so on. The Banana People group in fours, and their word for a group of four (a bunch) is *ba*. So the number seven would be rendered in Banana language as *ba-na-na-na*.

Naturally, those Banana tribesmen who wish to converse with the neighboring Hand People will have to translate between the two systems. A Banana merchant would become quite fluent in both systems—either by intentional study or (more likely) simply from constant daily use—and immediately upon seeing a *thump* will think to herself *ba-na*, perhaps even unconsciously.

Try counting to twelve in Banana language.

Our third group is the Tree People. In this tribe, numbers are communicated not by words or gestures but in written form—scratches on bark, let's say. Their grouping size is seven, and their first six numbers signs are:

Now, I myself can imagine getting a little confused by the last two. That is, I don't really trust my own sixness perception. But let's suppose that the Tree People can handle it.

Such a notation system clearly cannot continue indefinitely, so naturally they reserve a special symbol for a group of seven, namely a tree: ⩒. For the Tree People the number seven is quite special and sacred, being a solid and reliable number with no obnoxious leftovers. Of course, any Tree People who wish to exchange information (or bananas) with the other tribes would need to be able to translate fluently among the three systems.

Just to be clear, I am choosing to mention this hypothetical (and admittedly rather far-fetched) period of human history in order to illustrate a point. My point is that the artful rearrangement of numerical information—in particular, *the translation among different grouping sizes*—is the soul and essence of arithmetic.

As you would expect, each of these three tribes has its own cultural norms and expectations, and grouping size is no exception. Growing up in a world where people group things in fours endows that number with a certain mystical and emotional significance. When, from your earliest memories, you hear people counting: *na, na-na, na-na-na, ba, ba-na, ba-na-na, ba-na-na-na, ba-ba*, the rhythm of it gets into your bloodstream and consequently *ba-ba* starts to feel like a very solid, safe, and understandable quantity, from which a number like *ba-ba-na-na* is an unpleasant departure.

This is exactly how most of us feel about a number like twelve, for instance. In fact, the word *twelve* comes from an Old Teutonic root meaning "two left." This is not due to any particular intrinsic property of the number twelve but simply because we are used to grouping things into piles of ten. And because we do that, even our names for numbers are built around groups of ten. When we say "forty-six," for instance, we are really saying "four groups and six leftovers."

This is unfortunate for me because if I want to tell you about numbers and their properties or about arithmetic and its history, the very English words I am using are already biased toward a particular grouping size. So there is an issue of perspective. (In fact, it reminds me a little of teaching perspective drawing. The drawings that I make to illustrate the mechanism of perspective are themselves perspective drawings!)

So I'll need you to be open to the idea that any grouping size is just as good as any other and to be as "multicultural" in this respect as you can. In particular, try to resist the temptation to convert everything into familiar (ten-centric English language) terms. See if you can really experience these tribal number systems as a native.

Let's imagine a simple trading scenario in which one of the Hand People has some tools to trade in return for some bananas. Suppose he feels that each tool is worth *thump-thump-clap-clap-clap* bananas. A Banana tribesman who was sufficiently fluent in Hand language would probably think to himself: "*thump* is *ba-na*. So *thump-thump-clap-clap-clap* is *ba-na, ba-na, na-na-na*. I can take those leftover *na*'s and call them *ba-na*, so I get *ba-ba-ba-na*." In fact, a *really* fluent translator would probably not even need to do any of that kind of thinking but would just know from experience what *thump-thump-clap-clap-clap* is in terms of *ba*'s and *na*'s. One would quite rapidly get used to certain special numbers—like *thump-thump-clap-clap* being simply *ba-ba-ba*—as reference points, just as we think of thirty-one as being one more than the solid number thirty.

Each number language will have its own set of "nice" numbers. These are the numbers that are especially easy to say, which in turn depends on what numbers you have bothered to give special names to. In particular, one's grouping size always has a special (and usually *short*) name.

This means that when translating between grouping sizes, it pays to have a few reference points already learned. If I were a Hand–Banana translator, I would make it my business to know that *thump* is *ba-na*, *ba-ba* is *thump-clap-clap-clap*, and that *ba-ba-ba* is *thump-thump-clap-clap*.

Are there any numbers that
are nice in both languages?

But what if one is *not* very fluent? How can we make the correct translations? Being humans, we devise clever strategies. In particular, we can use representations of various sorts. Let's take one of the simplest: Piles of Rocks.

Imagine that you are an apprentice translator and one of the Hand People has just offered the tools for the above-mentioned price of *thump-thump-clap-clap-clap*. (Boy, I wish we were in the same room so that we could actually make these sounds together. Could you just clap and thump when I write those words? It might actually help to get my point across. It's also pretty fun.)

Not being fluent, you simply lay out rocks according to what was just said: *thump-thump-clap-clap-clap*.

Here we can literally see each *thump* as a handful and the *claps* as individual leftovers. Or, if you prefer a tidier arrangement, you can arrange your rocks in rows:

In this scheme, each row represents a handful, or *thump*, with the last row being an incomplete row of leftovers.

Either way, you now have a representation of the number that you can actually see and touch. And being primates with primate brains, we sure do like seeing and touching! That's why we make those "busy box" toys for infants—so they can turn the crank, open the door, and look in the mirror. (There are plenty of adult versions as well.)

As humble as it may appear, the Piles of Rocks system is actually a very powerful calculating device. To use it, you simply rearrange the rocks into whatever different grouping sizes you wish. For example, our Hand-Banana translator (that is, *you*) can simply slide the rocks around to make a new pattern:

Now it is easy to see the new groups of size *ba*, and we see that we have a leftover as well. (Perhaps even more artfully, you could simply move one rock in the earlier diagram from the second row to the third.) So we can confidently tell our fellow Banana tribesmen that the offer is *ba-ba-ba-na*.

Piles of Rocks is the world's first calculator. Notice that the user does not need to have any special knowledge or skill, just the ability to put rocks in piles (and not lose any, I guess). The downside is having to find (or carry around) a bunch of rocks or other convenient objects.

Let's try another example. Suppose now that the Tree People also wish to trade for some Hand People tools. They have carved wooden beads to trade and will pay ⚡⚡⚡Ṿ beads for each tool. You, as a somewhat inexperienced Tree-Hand translator, decide to get out your Piles of Rocks:

Rearranging this into handfuls (crudely or cleverly, depending on your taste and skill) yields the arrangement:

thump - thump - thump - thump -
clap - clap - clap - clap

This is what is known as "doing arithmetic." You have taken numerical information presented in one form, and you have reorganized it into another. Congratulations. Arithmetic doesn't actually get much more complicated than this, so if rearranging piles of rocks into different-sized groups makes sense to you, then it's probably going to be pretty smooth sailing from here on out. Of course, there's the question of getting *really* good at it—but that's entirely up to you and how much you want to practice and play around with it.

Now, you may have noticed that the number in the preceding example was rather large as numbers go. In fact, it is the largest number conveniently sayable in the language of the Hand People. Of course, they could say (and try to hear and remember) a number like *thump-thump-thump-thump-thump-thump-thump-thump*, but this brings back the whole perception problem again, which was the motivation for grouping and naming in the first place. The natural thing to do when numbers get big would be to start grouping the groups.

We'll focus on the Banana People. I want to imagine that a few centuries have gone by and that they have picked up the idea of using tools from the Hand People and also the idea of writing from the Tree People.

While we're at it, let's say that an entire Banana Civilization has come into full flower and with it the need to communicate larger numbers and more complicated transactions. The idea of four as a special and satisfying quantity remains, however.

The new written system might look like this. For each left-over we make a curved stroke, ⟍ (this being, as we all know, the universal symbol for a single banana). For a bunch of four bananas, we use a four-sided figure, □. Thus, the number ten (*ba-ba-na-na*) would be written □□⟍⟍.

Naturally, after a certain point, the number of *ba*'s would become so large that we would want to start grouping *them* as well. How many should constitute a group of groups? This is one of the more important decisions one has to make when designing a numerical representation system.

You might think that there's no question about it; we chose four (*ba*) as our grouping size, so that's that. Of *course* we will group our groups into fours—as opposed to what? Grouping things into fours and then grouping our groups into sixes? That would be insane!

But it happens all the time. Inches are grouped into twelves to make feet, and then three feet make a yard. And the old British monetary system had twelve pence to the shilling and twenty shillings to the pound. Such mixed-base number systems have always existed and continue to exist. The Babylonians, for instance, grouped their tens into "big groups" of sixty; that is, six groups of ten was their grouping of groups. Still, if one had one's way, it seems simplest to pick a grouping size and stick with it.

So let's suppose that for the same reasons that the Banana

People group their bananas into bunches, they also want to group their bunches into bunches. This quantity (*ba-ba-ba-ba*) is called *la* and is written ⊞. This is a nice symbolic representation of a bunch of bunches:

$$
\begin{array}{cc}
\bigcirc\ \bigcirc & \bigcirc\ \bigcirc \\
\bigcirc\ \bigcirc & \bigcirc\ \bigcirc \\[1em]
\bigcirc\ \bigcirc & \bigcirc\ \bigcirc \\
\bigcirc\ \bigcirc & \bigcirc\ \bigcirc
\end{array}
$$

Now larger numbers can be more easily represented and compared. For example, the number that the Tree People write as ꘎꘎꘎ would be understood by the Banana People as ⊞◱\ (*la-ba-na*).

How would the Banana People say and write ꘎꘎꘎꘎?

These three tribal number languages are examples of what are called *marked-value systems*, meaning that each word or symbol has a fixed, definite meaning. No matter where the symbol ꘎ or ⊞ occurs, it always represents the same quantity. In particular, for representation systems like these, the order of the symbols doesn't matter. Just as the change in your pocket can jiggle around without affecting its value, the representations

$$
\text{⊞◱◱\\\\\\} \qquad \text{\\⊞\◱\◱} \qquad \text{\◱⊞\\◱}
$$

all refer to the exact same number. This is a very convenient feature and makes systems like these particularly easy to work with. Marked-value systems (such as money) are very sturdy; you can scramble up all the symbols on the page (or all the coins in the jar) and it won't affect any of the information. Of course, it will affect the *form* that the information is in, so if you wanted it organized in a particular way then you might want to be a little more careful.

In particular, most people like to arrange their numbers (and often their money as well) into a form that makes for easy comparison. Usually this is done by collecting together the largest denominations first, and then working your way down to the smallest amounts. In other words, even though we can write our symbols in any order, it pays to put them down *in order of importance*. This makes it easy to see what the big picture is: "Oh. I've got five hundred and something," as opposed to giving a penny and a twenty-dollar bill equal status.

Any pile of rocks on the table represents that number of things no matter how it is arranged, but some arrangements are more communicative than others. Even though the representations ⊞□□\\\ and \\⊞□\□ refer to the exact same number, the first one screams out the headline "*la* and then some," whereas the second one hides its largest quantity under a bushel of bananas.

So we will usually want to express our numbers this way, from largest to smallest (whether that is written from left to right or not is a cultural choice). Again, the point of doing arithmetic is for comparison, and expressing our quantities in this way makes comparison particularly convenient.

For the same reason, we usually prefer our number representations to be "packed up" into nice groups, rather than spilled out all over the place like so many socks and underwear. It can be confusing to compare packed and unpacked representations:

ㅤㅤㅤ⊞⊞□\\ㅤㅤㅤㅤ□□□□□
ㅤㅤㅤㅤㅤㅤㅤㅤㅤㅤㅤ□□□□□

Here the second number is actually larger, even though it has no *la*'s explicitly written. So although the second representation is perfectly meaningful and unambiguous, it is somewhat inconvenient for comparison purposes.

On the other hand, nothing could be simpler than to compare two numbers that are both nicely packed into groups and organized by size.

⊞⊞☐☐\\   ⊞⊞☐\\\

Here the first number is clearly larger: both representations contain the same amount of *la*'s in them but the first has more *ba*'s. This way of organizing and comparing is called *lexicographic ordering*. It is essentially the same idea as alphabetical order. First compare the largest groups; whichever number has more of them is the bigger one. If there is a tie, then compare the next grouping size down, and so on.

So, for the purposes of comparison, we will usually want to order our symbols in this fashion. There will be times, however, when it will be more convenient to keep things unpacked and unordered. The point is we can do whatever we want. If you organize things in such a way that it takes a little longer, or you have to redo or undo something that you have already done, so be it. The stakes are pretty low. Mostly the idea with arithmetic is to have some fun, keep track of a few things, and occasionally enjoy a bit of cleverness.

> *You are a Banana tribesman with la-la-la bananas*
> *to trade for tools and wooden beads. Each tool costs*
> *thump-thump-clap-clap-clap bananas, and*
> *a single banana is worth two wooden beads.*
> *After buying three tools (and eating one banana for lunch),*
> *how many beads (in Tree language) can you afford?*

EGYPT

One of the earliest and simplest examples of a marked-value system is the one used by the ancient Egyptians from about 3000 BC. The idea is exactly the same as with the (fictional) tribal systems we were just looking at, except that this one really exists in the historical record. Also, as with most numerical representation systems used by humans, the grouping size is *ten*.

The number ten occurs often as a grouping size, not for any intrinsic mathematical or aesthetic reason but simply because it is the number of fingers we happen to have. In fact, ten (as a number) does not possess any particularly attractive features. Twelve would be better for divisibility purposes; eight is smaller and well suited to repeated halving. Ten is a purely cultural choice, and if you ask me, it's a bit too large.

In particular, ten is well beyond the limit of most people's number perception abilities. So in order to design a convenient marked-value system, the Egyptians first needed to solve the perception problem.

The Egyptian symbol for a leftover was the usual stroke, I. Instead of the five-barred gate, however, the Egyptian idea was to use *stacking*. Rather than eight being represented as ЖIIII, the Egyptians preferred IIII. In general, they tried to avoid using more than four strokes in a row. Thus, the numbers one through eight were typically represented as:

I	II	III	IIII	¼	III	W'	IIII
one	two	three	four	five	six	seven	eight

Notice how the stacking patterns also make it easy to see if a number is even or odd—the stacks either line up or they don't. Now the number nine becomes something of a problem, since it would seem to require a row of five strokes.

Rather than allow this, the Egyptians had a different plan: nine is three rows of three.

|||

nine

So the stacking patterns allow us to tell at a glance what we've got. As with dice, the designs are simple enough to be easily learned and recognized. We even have some flexibility in our choice of stacking patterns. The Egyptians often wrote four as two on top of two, for instance.

Since our grouping size is ten, we will need a special symbol for it. The Egyptians chose ∩, which is thought to represent a heel mark (presumably coming from the measurement of plots of land by digging in one's heels every ten paces).

Naturally, we use the same stacking patterns for groups as we did with leftovers. Thus, the number that we call forty-five would be written ∩∩ ⦚. It is also perfectly fine to write it the other way, as ⦚ ∩∩, depending on whether you are writing right to left or left to right. The Egyptians did both.

Of course, as the numbers we are dealing with get larger—and civilization tends to make that happen, what with storing grain and building armies and raising taxes—we will need symbols of greater value to denote groups of groups and so on.

The Egyptian symbol for a group of groups (that is, one hundred) is ९, supposedly a coil of rope. For a group of ९'s (what we call one thousand) they used the lotus flower symbol, ⚘.

The Banana People have been conquered by the Egyptians and now they must learn to use the Egyptian system. How would these numbers be translated:
⊞⫝⫝\, ⊞⊞⫝\\\, ⊞⊞⊞⫝⫝\\ ?

Another wonderful innovation, which replaces Piles of Rocks as a calculation tool, was the introduction of *counting coins*. The idea is simply to make objects (usually wooden or ceramic chips of some kind) that are marked so as to indicate their value. These coins do not need to have any actual worth whatsoever, just as long as they *stand* for specific values. In a way, this is another level of abstraction: a coin marked ∩ represents a pile of ten rocks, which in turn represents whatever it is that you are actually counting.

Let's imagine that we have a large bag full of these counters, marked with the various symbols ı , ∩ , ℗ , and ⚱ . We can then spill them out onto the table (also known as the counter) and sort them and arrange them into piles or rows however we wish. This then becomes a simple and convenient calculating device, or *abacus*. An abacus is simply a manual representation system; that is, a way of representing numbers by things that can be held and manipulated. Just as with Piles of Rocks, counting coins can be grouped and rearranged easily, and calculating can be done relatively quickly. Then the results can be written down at the end, if desired.

Make your own set of Egyptian counting coins.

Of course, for very simple calculations, an abacus system is usually unnecessary; you can often do such computations in your head, rearranging symbols and keeping track of information as you go. This requires a decent memory, however, and many people throughout history have found this sort of thing annoying and mentally painful. On the other hand, lots of people really enjoy mental arithmetic and even consider the use of an abacus system to be cheating or in some way beneath their dignity. No matter. If you enjoy storing and moving around detailed information in your head, then please do. If you would prefer to use some sort of calculating device, then be my guest. Both approaches to arithmetic are fun and full of surprises and amusement.

Anyway, to use the Egyptian counting coins is simplicity itself. We merely translate the numbers directly, each written symbol being replaced by the corresponding marked coin, like so:

And now our number can be manipulated (literally!), set aside, or combined with other quantities easily.

Imagine you are an Egyptian scribe, circa 1850 BC. The Pharaoh has asked for an accurate accounting of the grain supplies. The three granaries contain the following amounts, measured in bushels:

Thebes Granary:	𓏲𓏲 𓎆 𓃻𓃻𓃻 𓎛𓎛𓎛 II
Giza Granary:	𓎆 𓎛𓎛𓎛 III
Aswan Granary:	𓏲𓏲𓏲 III

What is the total amount of grain?

Assuming you are not comfortable calculating this total in your head, or you don't want to risk making an error and incurring the wrath of the Pharaoh, you make your way to the counting house and get behind the counter. Taking out your sack of counting coins, you proceed to lay out the numbers carefully:

To calculate the total, you simply push the piles of coins together:

The great thing about marked-value abacus systems is their wonderful flexibility—it doesn't matter how you scramble them up or rearrange them, the coins will always represent the same quantity. (Later, we will see some more modern abacus systems that are far more delicate but have their own advantages.)

So that's it, you've done the computation! The total is exactly the amount sitting there on the table. There are a couple of problems with this representation, however. First, it is inconvenient to have to carry a bunch of coins around (and you could accidentally drop a few, which, if they happened to be lotus coins, would be disastrous). We would prefer a more portable written representation. Secondly, for the purposes of comparison (which will undoubtedly occur at some point), we would want the number packed up into tidy groups, not eighteen coils of rope in a row.

So the next step is to "cash in" or *exchange* the coins until they are as compact as possible. In particular, if we have a group of ten coins of the same kind, say heel marks ∩, we can trade them in for one coil of rope ꝯ. This clearly doesn't change the actual number we are talking about, because ꝯ literally means ten ∩'s. This exchange maneuver does however affect the *form* our representation takes, and that is in fact the entire point. Every number can be represented in a variety of ways, and we want to choose a form that is as useful and convenient as possible. Sometimes, happily, this means doing absolutely nothing. Other times, we may want to "clean up" the representation a bit. It all depends on the circumstances and what we want in the particular situation. In this case, where we are presenting a number that will be recorded and compared with other numbers, it is a good idea to pack it up and arrange it in order of size.

So we trade in ten coils for one lotus, and ten heel marks for one coil. That leaves us with four lotus coins, one coil of rope, two heels, and a pile of leftovers:

Trading in ten leftovers in exchange for a heel, we finally get:

We now have a compact representation of our total, and all that remains is to write it down: ꕥꕥꕥꕥ ꕯ ∩∩∩ I.

Notice that these exchanges can be made in any order and at any time. You have complete flexibility about if and when you want to do any of these trades. Sometimes you may perform a certain exchange, only to find yourself later making the exact opposite trade (e.g., in order to subtract). Part of being a fluent and efficient arithmetician is learning to avoid these sorts of redundancies. Think ahead and try not to do unnecessary actions—that is, if you want to. It's not really a big deal; it's just about the craft of it and how much you care about doing something well—just as with knitting.

In particular, it often saves time to do your exchanges "from the bottom up." That is, cash in your leftovers first, then the groups, then the groups of groups, and so on. This will prevent your having to go back and recash a pile that you have already dealt with.

*Three Egyptian goatherds decide to combine their flocks
for safety. The first goatherd has 𓏲𓏲 𓎆𓎆 𓏻 goats,
the second has 𓏲 𓎆𓎆𓎆 𓏲𓏲𓏲 , and the third has 𓏲 𓏲𓏲 .
Despite their precautions, the wolves take 𓎆𓎆𓎆 𓏲𓏲𓏲 .
How many goats remain?*

ROME

The most popular and familiar marked-value representation system is the one used by the ancient Romans. This is arguably the most successful arithmetic system in human history, at least in terms of longevity. It is simple, convenient, and easy to learn. The Roman system also provides a rather different solution to the problems of repetition and perception. The grouping size is still ten, and being a marked-value system, it has special names for a group, a group of groups, and so on. In the Roman system these are represented by various letters of the Roman alphabet. (The Greek and Hebrew alphabets were also used in a similar way.) Thus, leftovers were denoted by the letter I (no need to specify a capital I, since the Romans had no lowercase letters), a group of ten was represented by an X, and a group of groups by the letter C (the first letter of the Latin word *centum*, meaning "hundred"). A group of ten hundreds was written M (for *mille*, "thousand").

The new idea was to introduce *subgroup* symbols in place of stacking. Instead of a long string of Is, which is hard to read and interpret, the Romans created a new symbol, V, to stand for five Is. Thus, rather than an obnoxious sequence of seven identical letters IIIIIII, one would simply write VII. Of course, the same thing was also done with the larger grouping sizes: L means five Xs (that is, fifty), and D stands for five Cs (five hundred).

What we are doing here is making a trade. We are eliminating the need for stacking but at the cost of having more symbols to get used to. Once again, the new symbols (and their English equivalents) are:

I	V	X	L
one	five	ten	fifty

C	D	M
one hundred	five hundred	one thousand

It is a good idea to view the subgrouping symbols V, L, and D as less significant. They are mere "convenience" symbols, as opposed to the major grouping symbols I, X, C, and M.

In essence, the Roman system is really the same as the Egyptian system, except that instead of stacking we solve the perception problem by introducing additional subgroup symbols. The choice of five as the subgroup size is very natural; it is the number of fingers on each hand. Both the Hand People and users of the five-barred gate system would whole-heartedly approve. Presumably, a creature with eight manual protuberances (e.g., an octopus) would find *that* number to be a convenient group or subgroup size as well.

Let's say you are a Roman scribe at the time of Cleopatra (circa 45 BC), and it is your job to translate the records of the Egyptian treasuries into Latin. How would you write the number 𒐜𒐚 𒐓𒐇 𒐐𒐐 ||| in the Roman system?

Here we can simply substitute the corresponding symbols (thankfully, the grouping sizes are the same) and use the convenient subgrouping symbols when there are five or more of a particular hieroglyph:

$$\text{𒐜𒐚 𒐓𒐇 𒐐𒐐 |||} \Rightarrow \text{MMMDCLXXVIIII}$$

Notice that because there is no stacking, all of the symbols on the right have the same height. This is very convenient for inscriptions in stone (which the Romans were keen on), as well as for keeping lines of writing neat and even.

By the way, there is a popular misconception about Roman numerals that I would like to dispel. You may have seen (e.g., in chapter headings, dates, and on clocks and watches) the use of subtractive notation, where four is written as IV and nine as IX. The idea is that since I is smaller than V, we can write VI for six and IV for four without ambiguity. But in fact the Romans never did this, and for good reason. The main benefit of using a marked-value system is the convenience of having symbols whose values are fixed and independent of the order in which they are written. The second you start placing symbols in some relation to others that needs to be kept track of, you introduce a delicate fragility into the system. You also add an extra layer of possible confusion to the computation process. No one would ever choose to do such a thing, as it would add unnecessary complexity.

Numbers are used for other purposes besides counting, however. For example, numbers make very convenient *labels*. Chapter numbers and copyright dates are good examples of instances where we often use numbers but not for arithmetic purposes. No one adds up house numbers either. These are simply stamps or labels we put on things for reference purposes. The subtractive shorthand became popular in the thirteenth century as a way to make such labels as short as possible. This was seldom done by the Romans and certainly never (by *anyone*) for the purposes of doing arithmetic.

> *While perusing the treasury records, you discover a damaged papyrus scroll containing three entries from the Aswan Granary:*
>
> 𓏏𓏏 ꝑꝑꝑꝑ 𓎛𓎛𓎛 III   𓏏 ꝑꝑ ∩∩ I   ꝑꝑꝑ ∩∩ IIII
>
> *Unfortunately, their sum is missing. Can you translate the hieroglyphs into Roman numerals and calculate the total?*

A nice way to interpret the subgroup symbols (one that the Hand People would certainly appreciate) is to think of them as handfuls or, better yet, as an open palm with five fingers. In this way, a number like VIII can be viewed as a hand and three fingers:

VIII

This is why I like to think of V as being part of the I-family. Similarly, when I see DC it looks like six to me (of course, I have to remember that it is six *hundreds* that I'm counting). So a fluent user of the Roman system would look at a large, complicated number like

MMMDCCLXVIII

and break it up in her mind as MMM (three thousands), DCC (seven hundreds), LX (six tens), and then finally VIII (eight leftovers).

Of course, the Romans did not speak English (it didn't exist as a language back then) and had their own Latin names for numbers, many of which are related to our own English number words. For example, the above number would be read in Latin as *tres milia septingenti sexaginta octo* and in English as three thousand seven hundred sixty-eight.

Computation in the Roman system is almost as easy as it was in Egypt. Being a marked-value system, there are no worries about where we place the symbols (although the Romans consistently wrote them left to right, from larger to smaller). In particular, if we are calculating a total we can just gather all the symbols together and arrange them however we like.

Imagine you're a Roman merchant assigning cargo to be loaded onto a Mediterranean trading vessel. The ship can hold two thousand tons of cargo. The goods at the dock, together with their weights (in tons), are:

Wine: DCCLXVII
Olives: DLII
Togas: DCLXXIII

Can all of this fit on the ship?

For this calculation we can simply count the number of occurrences of each symbol and exchange when needed:

DDDCCCLLLXXXVIIIIIII ⇨ MDCCCCLXXXXII

And so all three shipments can be safely loaded on board with a little room to spare.

Just as with the Egyptian system, you may find yourself wanting some sort of an abacus to help you do your calculations more easily and with greater confidence. And in fact the Romans did have such a device, known as the *tabula* (Latin for "tablet"). Of course, you could still use counting coins, but the tabula is in many ways faster and easier to use. It also makes use of a powerful new idea: *place value*.

The idea is simple and elegant. We take a block of wood or marble and incise horizontal grooves in it, labeling each groove with the corresponding grouping and subgrouping symbols:

I like to make the subgrouping lines somewhat shorter, for reasons that will become evident. To represent a quantity on the tabula abacus, we simply place small stones in the grooves, one for each of the corresponding written symbols. Thus, the number MMDCCLXVII would be placed on the tabula as:

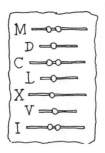

These stones were known as *calculi* (the plural of calculus, or "counting stone"). This is the Latin root of such words as *calculator* and *calculation*. In fact, the word *calculus* itself comes from *calx*, meaning "limestone." So there is an amusing etymological connection between the words *calculator* and *calcium* (also *chalk*).

The important thing here is that the calculi are not marked in any way; they are all identical. What gives a counting stone its value is where it is *located*. This is the place-value idea. A calculus stone sitting in a bowl has no meaning at all until you drop it into a groove. Then it takes on the corresponding value. That means you don't have to keep track of any markers or denominations. It doesn't really even matter what you use for calculi, as long as they fit in the grooves and stay put (I wouldn't recommend using ladybugs, for instance).

Make your own tabula abacus.
(If you like, you can draw lines on a piece of paper
and use buttons or pennies in place of stones.)

In moving from counting coins to the tabula, there is a tradeoff. We benefit from the flexibility of having unmarked calculi—a bunch of small, identical objects is easier to deal with than having to sort and keep track of marked denominations—but we pay a price. The price is a certain amount of *fragility*. Because it now matters where a counting stone is placed, we have to be careful not to accidentally move it. If the cat jumps onto a pile of marked-value counting coins, it's no big deal; their value is unaffected. But if Fluffy starts walking around on your tabula, a whole morning's worth of calculation could be ruined. So that's the trade: flexibility and convenience in exchange for being careful. This is a deal that has been made many times in the history of arithmetic, as we shall see.

Let's say you're a Roman scribe in charge of the olive oil storehouse. The senate has ordered MMDCXXXV flasks of oil for their upcoming bacchanal. Checking the supplies, you find MCCCLXIII flasks upstairs and another MCCLXXIII in the cellar. Do you have enough to fill the order?

Let's calculate the total using the tabula. We'll start by loading the first number (MCCCLXIII) onto the abacus, placing the calculus stones on the left side of each groove:

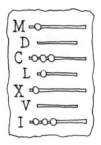

Now we can add in our second quantity (MCCLXXIII) on the right hand side:

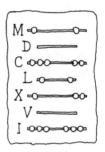

Our desired total is now sitting there on the tabula, only not in the most convenient form for comparison purposes. We'll need to do some exchanging to make it as simple and packed up as possible. Starting at the bottom, we can trade in five stones on the I-line for one on the V-line. Then the two stones on the L-line become one stone on the C-line. This gives us a total of six C-line stones, five of which can be cashed in for a stone on the D-line. After performing these exchange maneuvers, the tabula now looks like this:

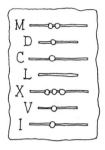

The total is now easy to write down: MMDCXXXVI. Comparing this to the quantity we need (MMDCXXXV), we see that we have exactly one flask of oil to spare. So the senators will enjoy their debauchery, and nobody will have to get stabbed or anything.

Notice how the exchange procedure is slightly more complicated on the tabula. That is another part of the trade. We replaced stacking by subgrouping, but then the exchange rate is different for groups and subgroups. This is why I made the subgrouping lines shorter. Five stones on a long line is worth one on the next higher (short) line, whereas it takes only *two* stones on a short line to cash in for a stone on the next higher (long) line. But that's not really so bad—at least it's the same consistent pattern for all long and short lines. In any case, it doesn't take very long to get used to it, and even small children can become quite fluent at using the tabula after a few hours of practice.

Use the tabula to calculate the sum of MCCLXVII, MDCLVI, *and* LXXXVIIII. *Is it more than* MMM?

I like how easy it is to read the stone patterns on the tabula. Numbers like seven or seventy appear as one calculus stone on top of two, which feels just like a hand and two fingers.

By the way, if you are interested in becoming skilled at the tabula (an important part of the education of a Roman scribe), you may wish to try a somewhat more advanced technique: rather than taking five calculi out of a long groove and replacing it with one on the next line, you can instead remove four of them and simply slide the fifth one up a line. This saves some time, but of course it also opens up a world of possibilities for confusion. You may find yourself staring at the stones in your hand with no clear idea of where you were in the exchange process, as it slowly dawns on you that you have *lost track*.

How would you design a tabula abacus
for the Banana People?

CHINA AND JAPAN

The Roman system is an interesting mix of both marked-value and place-value ideas. The written system is fully marked (including marked subgrouping symbols), whereas the abacus system—the tabula—is entirely place-valued. Another interesting example of this kind of hybrid system is the one used in China from about 200 BC up until the last century. The same written system was also widely used in Japan, having been imported from China (along with calligraphy and literary culture in general) around 500 AD.

The new idea here is a bold (and perhaps definitive) solution to the perception problem. Instead of subgrouping, where new symbols are introduced to cut down the amount of repetition, the Chinese system goes further and gives *every* number from one to nine its own special symbol. This is a pretty extreme solution. Now there is no repetition at all to worry about, but the cost is fairly high—we need to learn a *ton* of new symbols. These (along with their English equivalents) are:

一	二	三	四	五	六	七	八	九
one	two	three	four	five	six	seven	eight	nine

Notice that the first three symbols (apart from being horizontal and somewhat stylized) are essentially the same as those employed by the Egyptians and Romans. After that, though, it's an entirely new set of (rather arbitrary-looking) characters.

So, unlike the simple repetition-based representation schemes, the Chinese system has a bit of an initial hurdle: we need to memorize these nine symbols and what they stand for. On the other hand, most alphabets contain twenty-something letters, not to mention all the punctuation and accent marks we end up learning. So nine new symbols isn't so bad. And it does utterly solve the perception problem.

There is really no possibility of confusion once the symbols are learned.

Of course, we will still need grouping symbols as well. These are:

十　　　百　　　千

ten　　hundred　thousand

Here is the way the number MMDCLXXVIII would be rendered in Chinese:

二
千
六
百
七
十
八

First off, notice that the writing goes from top to bottom, instead of left to right. Second (and more important), notice how each grouping symbol is preceded by a single character that counts how many of those groups we want. So instead of repeating, as in MMM, we write

三
千

This can save a lot of time and space, as illustrated by these various representations of the number nineteen:

卌卌
卌IIII　　　∩ III　　　XVIIII　　　十
九

In order to read a Chinese or Japanese numeral, we simply look for the grouping symbols and see how many are indicated by the preceding character. It is also customary not to include such counting symbols when you have only one of a particular grouping size present. For example, here are three numbers written in both Roman and Chinese representations:

千		
三		九
百	六	百
十	百	四
四	二	十
MCCCXIIII	DCII	DCCCCXXXX

How would you write the following numbers in the Chinese system?

⊞⊞◻◻◻\\ , ⚮ ⵚ ⋔ Ⱶ ,
DCLV, *eight hundred ninety-one.*

As you might expect, the Chinese and Japanese cultures have devised their own unique abacus devices as well. The Japanese abacus, the *soroban*, is the simpler of the two, and is closely related to the Roman tabula. The new feature of the soroban is that instead of grooves containing pebbles, we have beads mounted on a bamboo frame:

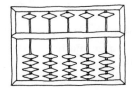

To use the soroban, place it flat on the table as shown so that the beads are free to move along the posts. To place a number on the soroban, we must first "clear" the abacus by pushing all the beads away from the crossbar. This is analogous to removing all the calculi from the tabula.

Each post of the soroban corresponds to a pair of grooves on the tabula. For example, the rightmost beads count the ones, or leftovers. The four beads below the crossbar represent the four potential calculi on the I-line of the tabula, and the single bead above the crossbar corresponds to a possible stone on the V-line. To indicate that you wish a bead to count, simply move it up toward the crossbar (or down, if it is a five bead). Here is the same number represented in both systems:

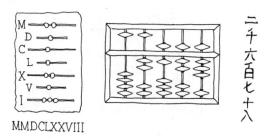

MMDCLXXVIII

The main difference between the Japanese soroban and the Roman tabula is that we don't need to carry around a bowl of calculi; the stones (in the form of beads) are built in. It's just a question of indicating which of them "count." So, rather than putting stones on and taking them off, we are simply sliding beads back and forth. The great advantage is *portability*. The disadvantage? Well, if you thought the tabula was delicate and subject to being bumped and disturbed by cats, the soroban makes the tabula look like the Rock of Gibraltar. Talk about fragile! One slight jostle and the beads go willy-nilly. So yes, you as a sixteenth-century Japanese accountant can put your entire abacus system in the pocket of your kimono, but you'd better find a calm, secluded place to do your figuring—preferably one with no pets or small children.

Make your own soroban.
(If you like, you can simply draw a frame on paper
and use pennies or buttons for beads.)

The other major drawback to the bead-frame abacus is that you cannot insert additional beads. On the tabula, you have the freedom to load a groove with as many calculus stones as you please—to be cashed in later, perhaps, but at least you have the *option* to leave a groove uncashed. With the soroban, however, you experience the elegant Japanese minimalist aesthetic of having no more beads than are absolutely necessary. In a way this is nice; your numbers are always displayed in fully packed form, plus it keeps the weight down. On the other hand—and this is a very big other hand—it means you have to do all the exchanging *in your head*. That's enough to make most people pretty squeamish, and it does mean that you have to practice a lot. Of course, anyone serious about doing a lot of arithmetic (e.g., our sixteenth-century Japanese accountant) will get quite a bit of practice on a daily basis and will gradually get used to it. Or you can go to school and get trained, whichever you prefer.

Let's say you're a medieval Japanese rice merchant. You have agreed to deliver sixty (六十) baskets of rice to the emperor's concubines. (I will write the Japanese numerals left to right to save space, and I'll throw in the concubines just to keep things interesting.) Checking the inventory, you find 二十七 baskets in the first storehouse and 三十五 in the second. Do you have enough rice to make the delivery? Will the concubines be pleased with you? (Now, *this* is why people do arithmetic!)

So we need to add the numbers 二十七 and 三十五 to see if they total at least 六十. Let's begin by loading the first number onto the soroban:

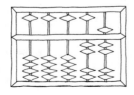

45

To do this, we slide up two ten beads (二十), and then enter the seven (七) as a five and two ones. (I'm using the words *up* and *down*, but since the soroban lies horizontally, it's really *away from* and *toward* the user.)

Now, to add in the number 三十五, we need to add three more tens and five ones. Starting with the ones, we would like to add five (五) by simply sliding down a five bead. Unfortunately, the five bead in the ones column is *already* down (meaning it counts). Here's where we have to get clever. In order to add five, we'll add *ten* and then take five away. This entails sliding up a ten bead and then sliding up the five bead in the ones column as well, so that it no longer counts. Our soroban now looks like this:

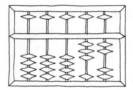

We just have to add in the three remaining tens (三十), and we're done. So we proceed to slide up three more ten beads, only to find that there are not enough beads remaining. It's time to be clever again. We want to add three beads to the tens column, so what we'll do is add *five* and remove two. Sliding down an upper bead and two lower beads does the trick, leaving us with:

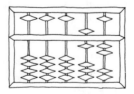

So our total is 六十二, and it turns out we have two baskets more than we need. (Do you think one of the concubines might like an extra basket?)

This is pretty much what it's like to use a soroban. It's actually a lot of fun, in its own way. You get to be clever and creative, and it also feels good sliding a bunch of beads around. You do have to be on your toes, however.

Of course, one way to alleviate the mental stress of the soroban would be to add more beads. The Chinese abacus, for instance, uses *five* beads below the crossbar and *two* above. This gives you a little more leeway (not having to exchange mentally quite so often) at the cost of some mathematical elegance. Going further, you could design an abacus frame with twenty or thirty beads on each post (mimicking the idea of a large bowlful of calculus stones), but then you're back to having perception difficulties: it's hard to tell the difference between a stack of seven beads as opposed to eight.

No matter what plan you come up with, it will have advantages and disadvantages, pros and cons, and will undoubtedly require a fair amount of practice before you get really comfortable.

What if the Banana People made a frame abacus?
What would it look like?

INDIA

Now that we've taken a look at a few different arithmetic systems—our imaginary tribal languages, the Egyptian system, and those used by the ancient Romans, Chinese, and Japanese—maybe it's a good idea to start thinking about what we want out of such a system. What makes one representation scheme better than another? What should we look for in an abacus? If we wanted to get all snobby and academic about it, I suppose we could even christen this sort of investigation "comparative arithmetic."

For symbolic numerical representation systems (that is, number languages), we've seen that an important requirement is that we solve the perception problem. There's not much point in coming up with a representation scheme if we can't easily tell at a glance what it is that we've represented. This means that our symbols need to be chosen with some care. They should be easy to tell apart and also quick and easy to write.

Most ancient Egyptians, for example, when doing a calculation would not bother spending five minutes carefully drawing perfect lotus symbols. Instead, they would use a shorthand version—quick, easy squiggles that get the job done. Often the number four was written as a single horizontal stroke, rather than spending the time to make four vertical ones. This sort of thing happens all the time with all forms of language and communication. We are lazy and easily bored, and we want to get things over with.

If our written system uses repetition, then of course we need to choose a grouping scheme of some kind (e.g., stacking or subgrouping) so that we can tell what we've got. The Chinese system solves this nicely, at the cost of memorizing a few more symbols. As for choosing a grouping size, we want to be a little careful. If our groups are too large then we will have a large number of potential leftovers and will either need to subgroup a lot or invent and learn busloads of new symbols.

The Babylonians, for instance, used a sexagesimal system, meaning that their fundamental grouping size was *sixty*. This means you might have as many as fifty-nine leftovers! Since no one wants to look at a string of fifty-nine identical cuneiform hash marks, some sort of solution to the perception problem is required. The Babylonians chose to subgroup in tens, and then use stacking patterns for the numbers one through nine—the point being that a large grouping size means having to take some extra trouble.

On the other hand, if your grouping size is too small, then although there are no problems caused by repetition, things get annoying when the numbers get big. The Banana system, for example, does a great job representing numbers on the order of a few dozen, but when the numbers get very large (say, in the millions), we're going to need tons of higher grouping symbols and the representations will be obnoxiously long.

The compromise we seem to have more or less settled on is to use *ten* as the standard grouping size (bakeries with their dozens and grosses notwithstanding). This is not so bad as far as it goes, but on the whole I'd say that ten is a bit too big. Maybe eight would be better. (Hopefully, the alien octopodes will one day conquer Earth and force us to use their elegant and well-designed octal system.)

Thus, our criteria for written systems should be ease of perception, ease of reading and writing, and a sensible grouping size.

What are the important features of an abacus device? A big one is portability. It's pretty annoying to have to carry around a heavy and unwieldy marble tabula, not to mention a sackful of calculus stones. It's much nicer to slip a light, self-contained bamboo soroban into your pocket and be on your way. (A more modern example would be using a laptop computer versus having to reserve time on the university mainframe.)

Another serious issue, as we've seen, is fragility (sensitivity to cats, in particular). That's a big tradeoff with the soroban, to be sure. Perhaps most important, there is the issue of *simplicity*. How easy is it to learn to use? Is it straightforward and user friendly, like Piles of Rocks and counting coins, or does it require a fair amount of training and mental gymnastics, like the soroban?

The truth is, there's no best solution. Every written and spoken number language and every abacus device ever invented has its advantages and disadvantages. Undoubtedly the best approach is to be so flexible and intelligent that you can easily move from one to the other. This of course would entail doing some work and would require that you care. If you do not, then the typical strategy would be to become reasonably proficient at the language and abacus system in use in your particular culture. Luckily for you, you happen to live in a golden age of information processing, and the handheld electronic calculator is easy to use, cheap, and ubiquitous. So lucky you.

On the other hand, we're not here just to be accountants. The point of studying arithmetic and its philosophy is not merely to get good at it but also to gain a larger perspective and to expand our worldview. At least that's why I enjoy learning things. Plus, it's fun. Anyway, as connoisseurs of arithmetic, we should always be questioning and critiquing, examining and playing.

All of which brings me to what I wanted to talk about: India. Specifically, the bold and original approach to arithmetic devised by the Hindu mathematicians of the sixth century AD. In some ways, this system combines the best features of the Roman and Chinese systems (though as far as I know there was no direct influence by either civilization).

Just as in the Chinese system, the Hindu number language uses distinct symbols instead of stacking or subgrouping. The grouping size (as usual) being ten, this entails choosing easily distinguishable written symbols for the numbers one through nine. In the traditional Devangari script, these are:

१	२	३	४	५	६	७	८	९
one	two	three	four	five	six	seven	eight	nine

In order to spare us having to learn an additional raft of new symbols, I'm going to instead replace these by their more familiar-looking (westernized) Arabic equivalents:

1 2 3 4 5 6 7 8 9

Before you start accusing me of historical inaccuracy and cultural insensitivity and so on, let me just point out that India is a big place, and different districts and principalities have used a variety of symbols and scripts over the centuries anyway. In any case, Arab traders were introduced to the Hindu system as early as 700 AD and quickly adapted it, using their own alphabet. So to avoid redundancy, allow me to explain the workings of the Hindu system using the more familiar Arabic number symbols. This historically important hybrid is commonly known as the Hindu-Arabic decimal place-value system.

The big new idea is this: instead of using a marked-value written system together with a place-value abacus (as both the Romans and Chinese did), the Hindu innovation was to make the written system place-valued from the start. To do this, we'll begin by drawing a frame of vertical lines:

This has the effect of dividing the page into columns. The idea is that each column corresponds to a different place value, like the long grooves in a tabula, or the posts of a soroban. Specifically, the rightmost column (or "place") will hold the leftovers, the next column represents the groups of ten, the next column hundreds, and so on. Thus, the number four thousand nine hundred seventy-two would be expressed as:

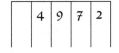

What we're doing here is using a place-value representation system with marked-value entries. The symbols in each column are known as *digits* (from the Latin *digitus*, meaning "finger"). This really is the best of both worlds, since the place values of the columns eliminate the need for grouping symbols like C and 百, while the marked values within the columns allow us to dispense with the rocks and beads. What this means is that our written system *is* an abacus system. And talk about portable—our abacus is a piece of paper and a pencil! There's nothing for the cat to mess up, either. So what is the price we pay for all this luxurious durable portability?

The tradeoff comes when we start calculating. Suppose we wanted to increase a number by one. On the tabula, all we would need to do is add a calculus stone to the I-line. On the soroban, we just slide a bead. But here, using the Hindu-Arabic pencil-and-paper abacus, we need to do something utterly new: we need to *change the symbol* in the leftovers column.

In the Roman and Japanese systems, the abacus is separate from the written encoding. We can perform computations (and any necessary rearrangements) using tangible objects and then write down the results later, after the smoke has cleared. But with the Hindu-Arabic system the writing and the calculating are inextricably linked. Instead of moving stones or sliding beads, our manipulations become transmutations of the symbols themselves. That means we need to *know* things. We need to know that one more than **2** is **3**, for instance. In other words, the price we pay is massive amounts of memorization.

To take a specific example, suppose we wanted to add two numbers together, say twenty-four and eighteen. If we were using a Roman tabula, we would simply place the two numbers XXIIII and XVIII on the abacus like so:

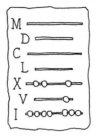

Cashing in five stones on the I-line for one on the V-line, and then exchanging two V's for one X, we easily obtain XXXXII, and we really didn't have to do too much thinking.

This same calculation on a soroban would be a bit more mentally taxing. We could start by loading the number 二十四 onto the soroban:

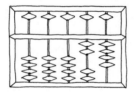

To add the number 十八 onto this, we would slide up another ten bead and then we would *want* to add eight by sliding down a five bead and sliding up three more one beads. Unfortunately, all four one beads are already up, so we'll have to be clever: we will add eight by adding ten and taking away two. So we slide up another ten bead and remove two ones, leaving us with 四十二.

With the Hindu–Arabic abacus, things get even more mentally strenuous. Now there's nothing to pick up or slide around at all. We'll just have to stare at these symbols and know how they "behave." This introduces a whole new level of abstraction. We replace concrete objects we can hold in our hands with abstract symbols dancing in our heads like sugarplum fairies.

The usual procedure is to put both numbers on the abacus, writing one above the other like so:

2	4
1	8

Then we can write their total below. Now, however, the entire exchange process must be done mentally. I will need to somehow know that four and eight combine to make one group and two leftovers. There are no rocks or beads to do it for me. Let's do the calculation in a couple of stages.

First, we can add the leftovers together: four and eight make twelve, which means one group and two leftovers. We can write that down on the next line. Next, we can move on to the groups of ten: two and one make three. We write that down as well. Then, collecting it all together, we get a grand total of four groups and two leftovers, as expected:

2	4	
1	8	
1	2	leftovers
3		groups
4	2	total

The point is, we need a way to get the totals in each column—not as a total number of rocks or beads but as a symbol that represents the total of two numbers represented by symbols. This is yet another level of abstraction added on to our already fairly abstract representation system. So when we say that 3 added to 5 makes 8, we are doing something one step removed from adding piles of rocks. In particular, we aren't counting. The symbol 3 isn't three of anything; it's a code that stands for three things. When we add we aren't pushing any piles together, we are manipulating and transmuting symbols so that they *encode* the result of pushing piles together, and that is a decidedly more abstract and mentally challenging activity.

The total of two or more numbers is usually written symbolically using the *plus sign* (+), a Renaissance-era abbreviation for "and." The *equals sign* (=) is used to indicate that the quantities on both sides of it are the same. Thus we write $3 + 5 = 8$, for instance, and the information is all there, short and sweet. Similarly, the *minus sign* (−) is used to denote subtraction. Thus we write $8 − 5 = 3$. People say, "three plus five is eight" and "eight minus five is three." (The words *plus* and *minus* are simply Latin for "more" and "less.")

Anyway, the upshot is that we need to memorize a whole bunch of symbolic sums and differences. In particular, we'll need to know (in symbolic form) the sums of each pair of numbers one through nine. One way to go would be to write all of this information down once and for all and keep it with us every time we want to do some arithmetic. This way, if 8 plus 4 happens to come up, we can just look on our chart and find that the total is 12. Of course, we would need a similar chart for subtraction as well. (And, in a pinch, we can always count on our fingers if we need to.)

> *Make a chart of all sums and differences*
> *of Hindu-Arabic digit symbols.*
> *Do you notice any patterns?*

Though this would spare us having to memorize a mass of symbolic data, it would quickly get tedious and annoying. It might be a reasonable way to get started, just as "Every Good Boy Does Fine" is a helpful mnemonic device for learning the lines of the treble clef, but such things have a tendency to quickly become obstacles instead of aids. When we learn to read there is a necessary initial stage of sounding out letters and fighting our way through a sentence, but what we are hoping to attain is *fluency*—the effortless comprehension of what is written without recourse to any external devices; the instant recognition of letter patterns seen hundreds of times before. The same goes for the Hindu-Arabic decimal place-value system.

The mistake that people usually make with this sort of thing (aside from trying to teach it to young children way before they are interested in it) is that they try to memorize by brute force. This is not really a very good way to memorize a body of information. I suppose that if for some reason you absolutely had to memorize the state capitals by tomorrow morning, then maybe you would simply have to cram them into your head somehow (and now that you know the capital of Kentucky is Frankfort, think of the whole new world of creative intellectual possibilities *that* opens up!).

No, it's no good remembering things that way, especially highly patterned interrelated information like the Hindu-Arabic digit sums. The best way to learn such things is to play with them a lot. The patterns will naturally become familiar just from experience. And you may find yourself forgetting them and having to reconstruct them from scratch. That is all to the good. The same thing happens with reading sheet music (I still screw up on the bass clef sometimes) and with spelling—that's what dictionaries are for.

In fact, getting stuck (say on $7 + 8$, for instance) is one of the best things that can happen to you because it gives you an opportunity to reinvent and to appreciate exactly what it is that you are doing: you are rearranging numerical information for comparison purposes. You have an amount of things, seven of them added to eight of them, and that is a complete and unambiguous description of what you have. There is no need to do anything. The number $7 + 8$ is not a problem requiring a solution, nor a question seeking an answer. It is a number, that's all. Oh, what's that you say? You want to compare it with something else? Well, all right. That's another matter. Now we might want to rearrange the form that our representation takes to make the comparison more convenient. We might need to know if that many eggs (one hen laid seven and the other eight, say) will all fit in a single egg carton—that is, whether seven added to eight is more than twelve.

Assuming we wish to operate using the Hindu-Arabic system, we would then want to express these numbers symbolically by grouping them into tens and leftovers. So the question then becomes: How many groups of ten do we create when we combine seven and eight, and how many leftovers?

The way I like to do this is to take one of the numbers, say eight, and ask myself what I would need to do to bring it up to a full group of ten—in this case, I'd need to increase it by two. Thinking in terms of rocks, I would steal two rocks away from the pile of seven and move them onto the pile of eight, thus making a group. That reduces the seven pile by two, making it a pile of five. So I have a group and five leftovers, otherwise known as fifteen. When we say that seven plus eight is fifteen what we are really saying is that seven combined with eight can be rearranged into a group of ten and five leftovers. In fact, the very word *fifteen* is simply an abbreviation for five and ten.

So it is vital (if you want to attain any kind of real arithmetic fluency) that you play around and get to be friends with small numbers and how they form groups of various sizes—especially tens, if you want to use the Hindu-Arabic system. There is actually a lot of fun and satisfaction to be had by noticing such things and seeking out clever ways to arrange and rearrange quantities. For example, if I wanted to total up a list of numbers like 7, 8, 4, 3, 2, and 5, it is mildly amusing to notice that the 7 combines with the 3 to make a group of ten, and the 8 goes along with its partner 2 to make another, leaving me with 4 and 5, which I happen to know make 9, so I get two groups and nine leftovers, or 29. The word *twenty-nine* reflects this grouping, since twenty is short for two tens.

Anyway, my point is that by playing around for a while the behavior of your new friends will start to become familiar. You don't need to intentionally set out to memorize anything. Unless, of course, you want to.

Design a Hindu-style system for the Banana People.
There are much fewer symbols now.
Can you learn the various digit sums?

Let's suppose that we know all the single-digit sums or that we have a table of them handy. Just to give you an idea of the flexibility and convenience of our new system, let's see if we can answer this question: Can I load my one thousand eight hundred and four pound elephant *and* my six hundred ninety-seven pound gorilla safely onto the freight elevator, which has a two thousand five hundred pound weight limit?

We'll start by entering our two numbers into the abacus frame:

		1	8		4
			6	9	7

Notice that our first number happens to have no tens and our second number has no thousands, so of course we leave those spaces blank.

Now we can start totaling the various quantities of tens, hundreds, and so on. Just to give you an idea of your options here, we could start with the hundreds. Knowing (by whatever means) that six and eight make fourteen (i.e., a group of ten and four leftovers), we see that we have fourteen hundreds. Similarly, seven and four make eleven (the seven steals three from the four to make a group, leaving one), so we have a total of eleven leftovers.

A perfectly reasonable thing to say at this point would be that our total is one thousand, fourteen hundred, ninety-eleven. This is not at all wrong or bad—confusing, perhaps, but in no way incorrect. It's just that it's still unpacked and inconvenient for comparison purposes. We could even write it this way if we wanted:

	1	8		4
		6	9	7
	1	14	9	11

This is very much like putting two numbers on a tabula and then gathering all the calculi on each line together, the point being that we still have some exchanging to do if we want it as packed up as possible. On the tabula or soroban this means moving some rocks and beads around, getting rid of some here, adding others there.

On the Hindu abacus it means reinterpreting our symbols. Instead of exchanging ten pennies for one dime, or removing ten calculi from the I-line and replacing them with one stone on the X-line, we need to simply *understand* ten in the ones column as being the same as one in the tens column. Similarly, ten in the hundreds column can be thought of as one in the thousands column.

So another way of writing the same sum would be:

	thousands	hundreds	tens	ones	
	1	8		4	
		6	9	7	
	1				thousands total
	1	4			hundreds total
			9		tens total
			1	1	ones total

Here I've simply totaled each column separately and entered these subtotals into the abacus. Instead of writing **14** in the hundreds column, however, I've reinterpreted this, exchanging the ten in the hundreds column for a one in the thousands. Same with the eleven ones. So this is yet another form of our number. Of course, it is still somewhat unpacked and spread all over the place in various rows and so forth, so we still have some work to do. Namely, we need to total these columns to get:

	thousands	hundreds	tens	ones	
	1	8		4	
		6	9	7	
	1				column totals with exchanges
	1	4			
			9		
			1	1	
	2	4			collected totals
		1			
				1	
	2	5		1	final total

Notice that the tens column overflows this time to form a group of ten tens (i.e., one hundred) with no leftover tens. So in the end we find a total of two thousand five hundred one pounds, putting us slightly over the weight limit. It's good that we figured this out *before* we loaded the animals into the freight elevator!

Now, you may have observed the preceding calculation with a mixture of dismay and disgust. Yes, it was lengthy. Yes, it was somewhat inefficient. That is to be expected. The first time you do *anything* it's usually a bit of a mess. And, in fact, over the past fifteen centuries there have been a number of improvements made to the system, both in notation as well as in the way we operate the abacus.

If you are neat and careful, and reasonably well organized, one simple improvement would be to eliminate the frame itself and simply line the digits up as you write them:

$$
\begin{array}{r}
18\ 4 \\
\underline{697}
\end{array}
$$

That saves a lot of time and energy and has a simpler visual appearance as well. There are problems with this scheme, of course, the most obvious being that it places an extra burden on the user to be tidy and to have fairly decent penmanship. If that's not you, then I would recommend sticking to the frame and its lines—you could use lined notebook paper sideways, for example.

The other big problem with removing the gridlines is possible ambiguity: if I write **2 8**, do I mean twenty-eight (and I'm just giving the **2** and the **8** plenty of room), or do I actually mean two hundred eight, or even two thousand eight, and the blank space matters? We can't have a written system that requires us to leave set amounts of space between our symbols; we're way to lazy and sloppy to stand for anything like that.

So the Hindu arithmeticians came up with a wonderful,

revolutionary, and hilarious solution: create a new symbol to stand for *nothing*. We simply expand our number language to include a "blank" or placeholder symbol, so we can tell if a column is intended to be empty (as opposed to just being a space between consecutive symbols).

I think it is ironic and clever to have a symbol that stands for nothing. Not only does it mean that we are adding a new symbol to our representation system, it also means that we are subtly adding a new *number* to our world of quantities: *zero*, the number of lemons you have when you haven't got any lemons at all. Clearly, this is not a quantity we have any trouble perceiving (I know when I'm broke and when I'm completely out of chocolate), and it's also not a terribly complicated or difficult amount to work with, either. Nothing from nothing leaves nothing, as the song says.

Of course, zero has no real practical utility as a number. Its function in the Hindu-Arabic system is to be a *placeholder*—its meaning is to give meaning to the other symbols by telling us where they stand in relation to each other. As I'm sure you know, the usual Hindu-Arabic symbol for zero is 0. (This can be confusing, since it looks so much like the letter O, but that's what you get for importing an Arabic symbol into an Anglo-Saxon language using a Roman alphabet.)

So now we really can dispense with the abacus frame altogether and just write our numbers freely, using the new zero symbol whenever we wish to indicate an empty column. Thus, we can write **28, 208,** or **2008** with no possibility of confusion.

While we're on the subject of zero, let me just say that I've never quite understood what all the fuss is about. For some reason, people seem to get all bent out of shape about "the invention of zero" as some kind of landmark event, not only in the history of arithmetic but in the development of civilization itself. I say *poppycock*. I would further go on to say *balderdash*. The zero idea allows us to dispense with the abacus frame, that's all. It's a good idea, I grant. But it's not zero that is the breakthrough concept, it's the idea of a symbolic place-value system—frame lines or no.

Yet another way to improve the Hindu-Arabic abacus system is to reduce the number of exchanges by always operating from right to left: we cash in all of the leftovers first, collecting together any groups of ten, and then proceed to the tens column from there. This way, once a column is fully cashed in, it will never need to be revisited. Here is our previous example, worked in this somewhat more efficient manner:

$$
\begin{array}{r}
1804 \\
\underline{697} \\
1
\end{array}
$$

Starting at the right (with the ones), we get four and seven make eleven, which is one group and one leftover, so we write down that leftover. Now, remembering (and this is the key) that I just made a group of ten, this means that I actually have one more ten then just the nine that are written, so that means I have ten tens total. (Notice that this is all going on in our heads.) So that makes one hundred and no leftover tens. That means there will be no tens at all when the smoke clears. So let's write down that fact:

$$
\begin{array}{r}
1804 \\
\underline{697} \\
01
\end{array}
$$

At this stage, we have just cashed in the tens to make one hundred, which means we have to *remember* that. Less writing, more thinking—those are the terms of the deal.

Next, we move on to the hundreds, and we see that we have eight and six makes fourteen of them—and oh yes, the one more we had to remember from cashing in the tens—so that's fifteen hundreds total. That means five and ten of them, or five hundreds and another ten hundreds, which is one thousand. So that's five leftover hundreds, which we quickly write down:

$$\begin{array}{r} 1804 \\ \underline{697} \\ 501 \end{array}$$

And again we need to be on our toes and remember that we just made an extra thousand. This, together with the one we already have on the abacus gives us two of them, so that's the total number of thousands and we're done:

$$\begin{array}{r} 1804 \\ \underline{697} \\ 2501 \end{array}$$

So that's quite a bit faster and requires much less writing. Of course, the price we pay is that it's much more demanding on our memory. But really, what is it that we need to remember?

Assuming we're only adding a couple of numbers together, the worst that can really happen is that one column "spills over" and makes an extra in the next column. That is, we only really need to remember whether there's an overflow or not. Is that really so hard?

One way I like to think of it is to imagine that I'm sewing with an imaginary needle (we won't need any imaginary thread). I start by passing the needle through the ones column (from top to bottom), picking up the quantities contained in that column:

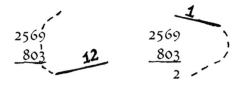

Thus the needle first picks up the nine, then the three, making twelve. Then I drop off the leftovers (2 in this case) in that column, leaving the 1 that may still be there due to making a group of ten. Then the needle goes back up to the top of the next column, and the 1 on the needle now gets counted as part of that column. (This cleverly takes care of the exchanging and reinterpreting process, because the 1 which formerly stood for ten in the ones column is now simply a one in the tens column.)

This process is usually called *carrying*, and here we are literally carrying the 1 on our imaginary needle (OK, maybe figuratively). What I then like to do is to make the carried 1 "active" by having it increment the topmost number in that column. So here the 6 becomes (in my mind) a 7 symbol. Then the process repeats until I finish with the last (leftmost) column.

The point being that all I need to remember and visualize is whether at the time of each "stitch" my needle has a 1 on it or not. This metaphor also emphasizes the somewhat mechanical quality of the Hindu-Arabic abacus, making it feel a lot like knitting and sewing and other relaxing manual crafts. Anyway, it's how I like to think of it. The good news is that it's pretty quick, not too mentally taxing, and it gets the job done.

Of course, there are those for whom even this small expenditure of mental energy is too much; who for some reason do not trust themselves to remember what's on their needle. In fact, many people have been trained to actually write down little 1's above the relevant column so they don't forget:

$$
\begin{array}{r}
{\scriptstyle 1\ \ 1} \\
2569 \\
803 \\
\hline
3372
\end{array}
$$

This has always struck me as a bit silly. How hard is it to remember that you just made a group of ten, just a second ago? Of course you should do as you wish. But if you want to attain real fluency, you'll want to dispense with the little 1's. Plus, they mess up your paper and obscure the original numbers. My advice would be to use a different abacus system, like a tabula or counting coins. Then you don't have to remember anything (except the exchange rate).

Naturally, things get a bit more demanding when you're adding three or more numbers together. For example, if we wanted the total of 3278, 867, and 2389, we would first align them as usual:

$$3278$$
$$867$$
$$\underline{2389}$$

Then the first pass of the needle would pick up 8, 7, and 9, for a total of 24 leftovers. (Notice all the mental carrying we need to do even at this stage.) Dropping off the four ones, we are left with a 2 on our needle for the second stitch:

$$3278$$
$$867$$
$$\underline{2389}$$
$$4$$

The 2 on the needle turns the 7 into a 9, which together with 6 and 8 makes 23, so we write down the 3 and leave the 2 on the needle:

$$3278$$
$$867$$
$$\underline{2389}$$
$$34$$

The carried 2 together with the 2, 8, and 3 in the third column give us 15, so we drop off the 5 and carry the 1 over to the final column, giving us a total of 6 in the thousands place:

$$3278$$
$$867$$
$$\underline{2389}$$
$$6534$$

And we're done. The same technique applies no matter how many numbers you wish to add together; the only real difficulty is keeping track of the column totals in your head. Whether or not this is something you want to take the time to get good at, it is nice to know that the method is perfectly general.

Of course, many people (including myself) often dispense with any systematic procedures whatsoever and simply rearrange and exchange mentally: "Let's see here, 2569 + 803. Two thousand five hundred and another eight hundred, that's thirty-three hundred, then I've got sixty and twelve, so seventy-two, making thirty-three seventy two, 3372. Done." Naturally this sort of thing requires a bit more concentration and a fair amount of experience, but it's well within most people's ability.

Sometimes, for instance, I prefer to move from left to right, despite the additional labor of recashing a column, because a lot of times I'll see that my total is already too big, and then I don't need to continue. Anyway, the point is that you can do whatever you like, and you should. Especially when it comes to inventing your own personal ways of operating.

You are a tenth-century Arab trader on the Mediterranean. One of your ships can hold 800 tons of cargo, and the other can hold 825 tons. Your shipments are:

Spices: 152 tons
Carpets: 721 tons
Tea: 312 tons
Silk: 465 tons

Can you fit everything onto the ships?

Now let's see what subtraction looks like in the Hindu-Arabic system. Right away this means we have a whole new boatload of symbolic "number facts" to memorize (e.g., **12** take away **7** leaves **5**). Again, the best way to learn such things is not to try; just work them out with rocks or on your fingers each time, and they will start to sink in after a while. Not that it's a big deal if they don't. I'm sure there are certain words you have to look up every time you read them and no matter what you do you can't seem to remember what they mean. I have that with the word *ontological* for some reason. (Damn! Now I have to go look it up again.)

So let's suppose we know all the small number differences from experience (all of this would be so easy if our grouping size were five or six, but alas). Now the question is, what does the mental exchange process look like when we subtract?

Imagine we are twelfth-century Arab silk merchants. We recently sailed to Naples with **1876** bolts of saffron and **1422** bolts of indigo dyed silk. We sold **1551** bolts of the saffron colored silk and **973** of the indigo. What is our remaining stock?

Let's start with the saffron-colored silk. Of course, most experienced merchants would do such calculations in their heads, but let's say we don't trust ourselves to do that. The usual procedure would be to write down the amounts, the smaller one below the larger (not that it really matters):

$$1876$$
$$\underline{1551}$$

On the tabula we would simply remove the requisite calculi: take one stone off the I-line, one from each of the L- and D-lines, and one from the M-line. Now, however, this must be done purely symbolically, which means by memory:

$$1876$$
$$\underline{1551}$$
$$325$$

If we happen to know the differences between small numbers, then we can simply write them down: 6 minus 1 is 5, 7 minus 5 is 2, and 8 minus 5 is 3. So we have 325 remaining bolts of saffron silk.

As for the indigo-colored silk, things are a bit more complicated. Placing the numbers on the abacus (i.e., writing them down, aligned on top of each other), we have:

$$
\begin{array}{r}
1422 \\
\underline{973}
\end{array}
$$

If we were using a tabula or a soroban, we would need to start exchanging. For example, on the I-line we would have only two stones and we need to remove three. So we would exchange one of the calculi on the X-line for ten stones on the I-line (or, if you wanted to save time, one stone on the V-line and five on the I-line). Now we would have enough stones on the I-line to allow us to remove the three we want to take away. So on the tabula, subtraction entails uncashing or unpacking some of the information to provide a supply of lower-valued objects for removal—while, of course, not actually changing the represented quantity, only its outward form.

With the soroban, we would need to do a similar, though somewhat more elaborate maneuver: we remove a ten bead (that is, we move it away from the crossbar so that it no longer counts) and then note (mentally) that since we need to put the ten back in, and we are also subtracting three, this is the same as adding seven, so we would add on a five bead and two one beads.

In the Hindu-Arabic system all of this exchanging must be done in our heads: we take one of our tens (so we now have only one ten left), and we crumble it into ten ones (so now we have twelve in the ones column. Taking away three leaves us with nine ones. So we can write:

$$
\begin{array}{r}
1422 \\
\underline{973} \\
9
\end{array}
$$

The only slightly confusing thing being that the 2 in the tens column is actually a 1, since we cashed in one of our tens already. (This procedure is commonly known as *borrowing*, but it's really more of an exchange than a loan; in any case, we've already paid it back.)

Some people go so far as to actually *cross out* the 2 and replace it with a 1:

$$
\begin{array}{r}
1 \\
14\cancel{2}2 \\
\underline{\cancel{973}} \\
9
\end{array}
$$

This is done out of fear, I suppose, that they will forget about the exchange that they just did two seconds ago. That seems ridiculous to me. First of all, it slows us down for no reason, and even more heinously, it messes up the very numbers we are interested in working with! Is it really so hard to remember what you just did? If you must do such things, why not simply place a dot over the number you borrowed from, or some other harmless mark to indicate that it is actually one less than it appears?

My advice is to shun such pointless and time-consuming devices and simply get on with the calculation. Just as we needed to remember whether our needle was "loaded" or not, here we just need to be aware of when we have stolen or borrowed, or whatever you want to call it.

So now we have only one ten and we need to remove seven. This will require another exchange. Taking one of our hundreds and exchanging it for ten tens (which is, after all, what a hundred is—our special name for a group of groups) gives us eleven tens to work with. Now we can easily take away the seven tens, leaving us with four of them:

$$1\overset{4}{4}22$$
$$\underline{9\overset{}{7}3}$$
$$49$$

The computation is going very smoothly—we just need to be conscious of the fact that the 4 in the hundreds column should be regarded as a 3. And again, since nine hundreds are to be subtracted, we still need to exchange once more: we turn the thousand into ten hundreds to give us thirteen of them, then remove nine, leaving us with four:

$$1\overset{4}{4}22$$
$$\underline{9\overset{}{7}3}$$
$$449$$

As complicated as this may seem at first—memorizing a bunch of symbolic information, exchanging mentally, remembering when we have done so, and aligning everything properly—with a little practice it becomes quite easy and even enjoyable. Like I said before, it's really just symbol knitting: a small number of basic moves that are repeated over and over to produce something of value—information, in this case. Now we know that we have **449** bolts of indigo cloth left, so we know that we cannot promise to deliver **450**, for instance.

*A warehouse fire has destroyed **288** of the **449** bolts of indigo silk. Moths have ruined another **75**. Do you still have enough left to fill the sultan's order of **84** bolts?*

One important thing to notice about the Hindu-Arabic system is that the exchange rate is *uniform*. Each column is worth exactly ten times the one next to it. There are no subgroups as with the tabula; each exchange is always the same: we take one away from a column and increase the column to the right of it by ten. Once this becomes familiar, then it ceases to matter whether you are working in the tens column or in the ten millions column; the process is always the same. That's a very nice feature.

There is one case where things can get slightly confusing, and that is when you are subtracting from a column and you need to borrow, but the next column is *empty*. That is, there's nothing there to cash in. For example, suppose we are faced with a subtraction like this:

$$
\begin{array}{r}
2057 \\
386 \\
\hline
1
\end{array}
$$

Here we are in the second column (the tens), having only five and needing to take away eight. The usual thing to do would be to cash in a hundred, giving us fifteen tens. But alas, we have no hundreds at all.

What a Roman would do, of course, is to take a stone off the M-line and exchange it for a bunch of hundreds so that there would be enough of them to use. For us this would mean regarding the 0 in the hundreds column as actually being a ten, and the 2 in the thousands column becomes a 1 in our minds:

2057 on paper ⇨ 1 10 5 7 mentally.

Now there are plenty of tens, so we can easily cash one of them in to make ten ones:

1 10 5 7 ⇨ 1 9 15 7

In essence, we are thinking of the number two thousand fifty-seven as being one thousand nine hundred fifteenty-seven. So when zeros are present there can be a drop more mental labor involved. Mostly, the effort is to keep clear about what you are doing and not get confused. But that's true with ordinary knitting as well; occasionally you drop a stitch and have to backtrack a bit. It's all part of the fun of it, really. So on paper, our calculation would now look like this:

$$
\begin{array}{r}
2057 \\
\underline{386} \\
71
\end{array}
$$

Here is where we gave ourselves the extra tens (so that we have fifteen of them), took away eight to leave us with seven, and now we just need to be awake to the fact that we are left with only one in the thousands column but nine in the hundreds. So we conclude with:

$$
\begin{array}{r}
2057 \\
\underline{386} \\
1671
\end{array}
$$

One thing I like to do at this point is to quickly add the two bottom numbers to make sure they total to the top number, as they should. Since I'm a lot faster at addition, this is an easy way to check myself. Sure enough, it all works out.

Generally speaking, when one is subtracting from a number it is always nice when its digits are large, because that cuts down on the number of exchanges. I usually go so far as to allow numbers in the teens and even twenties to occupy a single mental column. For example, if I need to add two numbers, say **453** and **866**, and then subtract another, say **395**, I usually won't bother cashing in at all: **453** and **866** is **12 11 9** (that is, twelve hundred eleven-ty-nine), and then it is easy to remove **395** to get **924**. Anyway, *real* fluency means you are comfortable with any sort of representation, and you often like to invent your own methods. So do that.

Which is larger, **5003** − **2684** *or* **4086** − **1767** ?

I guess my real point here (and with this book in general) is that there are many good strategies for encoding and manipulating numerical information, and you can use them in any way you see fit. Instead of thinking in terms of systems and rules, think of it more as options and tools at your disposal. There's no rule saying I can't have forty-seven in the tens column if I want to; the only issue is whether I know what I mean and what I want. So play around!

One more thing about the Hindu-Arabic system: now that we have a fully symbolic place-value representation system (which also functions as a lightweight and durable abacus), it is almost effortless to extend it to larger and larger quantities. Whereas the Egyptians, Romans, and Japanese would need to add new symbols at each new stage of grouping, the Hindu-Arabic system carries with it no such burden. The only thing needed is to add more *columns*, and that's trivial. In principle, we could even work with such astronomical numbers as

$$180224600381257928805.$$

It is amusing to note that our system allows us to easily write and calculate with such vast quantities, regardless of whether we can conceive or comprehend them.

There remains the question of how to *say* such a number. In a way, this presents us with a new kind of perception problem. Of course it doesn't really matter very much, since the digit sequence itself holds all the information. There is actually no real need for words like *ten*, *hundred*, or *thousand*. We could just say "three-two-eight" instead of "three hundred twenty-eight" (and people often do).

In fact, there are words out there, like *million*, *billion*, and *trillion* (and even absurd pseudo-words like *octillion*), that correspond to increasingly larger place values. In English, the usual practice is to break the columns into groups of three (notice how often we group things!) and invent a new grouping name every third place. Thus, a number like **40261396** would be separated as **40 261 396** and read as "forty *million,* two hundred sixty-one *thousand,* three hundred ninety-six." It has been customary (at least in the United States) to separate these groups by commas—that is, to write **40,261,396**—but that tradition seems to be on the way out. In Europe, the period (**40.261.396**) is often used as a separator. Curiously, in many Asian countries it is customary to introduce new place names every *four* columns.

> *The alien octopodes have arrived and have instituted their octal (base-eight) regime. Thankfully, they have allowed us to retain the digit symbols* **01234567**. *How would the number two hundred seventy-three be written in this new system?*

EUROPE

Now that we have arrived at a point in our brief pseudo-history of arithmetic where things are starting to feel a bit more familiar, I think it is especially vital to be conscious and clearheaded. The problem with familiarity is not so much that it breeds contempt but that it breeds *loss of perspective*. Having grown up with the Hindu-Arabic decimal place-value system and being constantly inundated with these particular symbols and digit sequences, as well as our names for them, it is easy to lose sight of the big picture and to allow convention (not to mention schooling) to substitute for understanding.

In particular, I want to make sure that we remain awake to the distinction between a number in and of itself (that is, a quantity in the abstract) as opposed to a culturally determined choice of representation. There is nothing sacrosanct about this Hindu-Arabic symbolic encoding of ours. It's one of many such systems that people have come up with, and though it is certainly widespread and popular, it is certainly not the only one in common use. The five-barred gate is still going strong, for instance.

Numbers have an unlimited variety of ways in which they can be represented, and regardless of how we may feel about the pros and cons of these various choices, I can tell you one thing for sure: *the numbers themselves don't care.* Six has no interest in your little pet name for it, or by what silly scrawl you care to represent it. Six is *six*, or rather, six is the entity that embodies sixness. Six is even, and six is one more than five. These are intrinsic properties independent of language and culture. The observation that six looks like a nine upside down is not really a statement about the number but about the shapes of some Arabic squiggles. The more you can step back and free yourself from the language (especially the choice of ten as a grouping size), the more flexible and mathematical your viewpoint will be.

As a mathematician, I tend not to think of numbers symbolically, or even necessarily as quantities. To me, numbers are *creatures* that exhibit behavior, and I occupy my time observing, studying, and trying to understand that behavior. Depending on the circumstances, I may choose to represent a number symbolically, but my choice will be guided by my own purposes and aesthetic sensibilities more than what the shopkeepers of my era happen to be using. All I'm trying to say is that we need to be wary and not let our familiarity with a particular system blind us to its arbitrariness.

The Hindu-Arabic decimal place-value system was introduced to Europe in the early thirteenth century. Instrumental in the spread and popularization of this new arithmetic system was *Liber Abaci* (Book of the Abacus), written by the mathematician Leonardo of Pisa, also known as Fibonacci. Essentially, this was the first European arithmetic textbook.

Despite its obvious advantages over the clumsier Roman system, which were immediately recognized by scholars and professional accountants, the Hindu-Arabic system was slow to catch on with the general public. Even as late as the eighteenth century, well-educated adults found it confusing and overly technical.

Eventually, the convenience of a symbolic calculating system free of rocks, beads, and coins, together with the increasing availability of inexpensive paper, finally won out over people's natural unwillingness to learn something new. (I wonder how reluctant people would be *now* to adopt any further improvements to the system?)

Actually, one such improvement was proposed following the French revolution of 1789. The new government, in its zeal to do away with all remnants of the *ancien regime*, voted to abolish the old system of weights and measures that had been in place since the time of the Romans, and replace it with a more modern, rational, and scientifically enlightened version, commonly known as the *metric system*.

The idea is that if we are going to adopt a numerical representation scheme that is based on a grouping size of ten—and certainly by the late eighteenth century the Hindu-Arabic system was the conventional choice throughout most of Europe—then it makes good sense to have all of our measuring units organized along similar lines.

Thus, instead of a mile being eight furlongs, each furlong being two hundred twenty yards, a yard being three feet, and a foot being twelve inches (which are then further subdivided into quarters, eighths, and sixteenths), the revolutionary government took the suggestion of the leading scientists of the day to adopt a more decimally oriented system of units: a kilometer is ten hectometers, each of which is ten decameters, a decameter being ten meters. A meter is then divided into ten decimeters, each of which consists of ten centimeters, which are then divided into ten millimeters. The way the units are grouped and subdivided then matches the grouping size of the number system. It's not the number ten itself that matters so much, it's *consistency*. We could just as well choose eight as our grouping size for both numerical representation as well as for our measuring units, and things would be every bit as convenient, if not more so.

Mind you, not all of these metric units are equally popular or useful in actual practice. Meters and kilometers are used quite frequently, hectometers almost never. Most people seem to prefer thinking of a half meter as fifty centimeters rather than five decimeters. Taste and habit certainly play a role in such decisions, but no matter what your preferences, it is always more convenient to have consistent agreement between your measuring units and your number system.

One interesting example where the old nondecimal units have been retained is in the measurement of time. For whatever reason, we have held on to the division of the day into twenty-four segments (known as hours), and the hour into sixty minutes, which are then further divided into sixty seconds. Clearly this goes back to the Babylonians with their standard grouping size of sixty. Perhaps this way of measuring time is simply too deeply ingrained. Changing ells and fathoms to meters is one thing, but noon not being twelve o'clock—madness! Maybe the French scientists feared that messing with the way people tell time would start a riot or some other violent protest. (The guillotines were still standing, after all.)

Nevertheless, there is in fact nothing special about twelve, twenty-four, sixty, or any other such grouping sizes. We could just as easily call a day ten periods (or whatever word you like) and subdivide these into tenths, and so on. The choice of twelve is a cultural and historical one, based on the fact that there happen to be twelve more or less equally spaced clumps of stars observable in the Egyptian night sky. One could keep track of time in the ancient world by seeing which constellation was overhead. So it's mildly annoying that my ten fingers conflict with my twelve constellations. In any case, the twenty-four-hour day and the sixty-minute hour seem to be fairly well entrenched. (Sometimes I even confuse the two systems and think that $4.59 is only a penny away from five dollars.)

The point is that Earth turns on its axis and it takes a certain amount of time to do that. How you choose to chop that up into smaller amounts of time is up to you. What is not up to you would be things like the number of days in a year or how many days there are between full moons. These numbers are built into the solar system and are in no way culturally determined. And in fact, both of these are somewhat unpleasant (and fractional), a year being just a hair under 365¼ days, for instance. (The Babylonian astronomers must have been bitterly disappointed that it wasn't exactly 360—six sixties!)

Incidentally, one holdover from the Babylonian system is the division of a full turn into 360 degrees of arc (which are often subdivided into minutes and seconds just as with hours). Though mathematicians long ago abandoned such a ridiculous and arbitrary scheme, it remains as the conventional angle measurement system among carpenters, architects, and even many engineers and scientists.

Again, the point here is consistency. If you are going to go around grouping things or breaking them up into pieces, it is far more convenient to choose a fixed grouping size and stick with it, as opposed to the haphazard sort of bundling and subdividing that seems to occur historically.

Rather than illustrate the convenience of a fixed-base measurement system, I think the case will be made even more strongly by subjecting ourselves to the alternative—a clunky and awkward mixed-base system of the Old World variety. And there is no such example quite as annoying and unnecessarily complicated as the British monetary system of the nineteenth century.

Given its history, it is no surprise that the traditional system of British units is a disaster. In pretty much every arena of measurement, from acres to ounces, from leagues to hogsheads, is written the record of scores of invasions and conquests, together with their inevitable cultural and linguistic side effects. The monetary system is merely one of a vast number of such casualties.

Two thousand years ago the Romans introduced to the British Isles their monetary system based on the *libra*, or "pound" of silver. Thus, the British pound sterling (still denoted by a fancy script £) represents the value of a pound (that is, sixteen ounces) of that precious metal. (This is also the source of the abbreviation lb. for pound.) Naturally, this is divided into twenty shillings (represented by *s.* for *solidus*, an old Roman coin). And of course, each shilling was itself divided into twelve pennies (the abbreviation for penny being *d.* from *denarius*, as you might expect). Thus, twelve pennies (or pence, as the plural is commonly rendered) make a shilling, while it's twenty shillings to the pound.

Putting aside the absurd profusion of denominations available to Jane Austen and her contemporaries (e.g., the *crown* worth five shillings or the *guinea* worth twenty-one shillings), we already have a logistical and computational nightmare just dealing with pounds, shillings, and pence.

Suppose you are a London shopkeeper in 1820. Lady Smythington-Jones has just run up the following bill of purchases:

pearl inlaid snuffbox, 1£ 8s. 6d.
set of six matched spoons, 13s. 8d.
two salt cellars, 2s. 7d. each.
pair of silver candlesticks, 1£ 14s. 4d.

What is her total, and what is her change when she hands you a five-pound bank note?

This is exactly the sort of arithmetic calculation that British schoolchildren would practice in their lessons. (Perhaps this example may help you to sympathize with the plight of Bob Cratchit and his fellow clerks, scratching their quills along columns of figures in the dim candlelight, enduring the tyranny of both Ebenezer Scrooge and a mixed-base representation system, even as Britannia rules the waves and the sun never sets on the empire.)

Organizing pounds, shillings, and pence into appropriately labeled columns (and doubling the price of the salt cellars in our heads), we proceed to produce the following receipt:

£	s.	d.
1	8	6
	13	8
	5	2
1	14	4

Next, we total these amounts in the customary way, starting with the pence (keeping in mind what every British schoolboy knows, that twelve pennies doth a shilling make). Adding in our heads (or on our fingers and toes if necessary), we find a total of twenty pence, or one shilling and eightpence.

Dropping off the eight pennies in the pence column, we proceed to tally up the shillings (having learned our lesson well, that twenty shillings make a pound). Together with the one shilling carried over from the pence column, we get forty-one. (Notice all the base-ten figuring we must do, even when ultimately grouping by twelves and twenties. What a mess!) This is of course the same as two pounds one shilling, so we arrive at a total bill for four pounds one shilling and eightpence.

£	s.	d.
1	8	6
	13	8
	5	2
1	14	4
4	1	8

Being slightly over four pounds, we will clearly be giving Lady Smythington-Jones a little under a pound in change from her five-pound note. If not for the eight pennies, it would be a simple matter of nineteen shillings. As it is, we need to reduce that by eightpence, so her change becomes eighteen shillings and fourpence. "And good day to you, milady" (tipping your hat ever so slightly).

The British finally abandoned this absurd system in 1971 in favor of a decimal currency. There are now 100 pennies to the pound, instead of the 240 in the former system. I must confess to having felt a certain sadness and loss of romance as £1 8s. 6d. ("one pound eight and six") gave way to the more prosaic, if undeniably more convenient £1.425. (I suppose I can drown my sorrows with a pint of lager from a stout publican who stands six foot one and weighs all of fifteen stone.)

While we're on the subject of consistency, I want to mention a useful and convenient extension of the Hindu-Arabic system that is frequently employed (and for some reason seems to generate a fair amount of confusion and dismay). Just as it makes sense to choose a grouping size and stick with it—bundling the same way at every level, so to speak—it also pays to subdivide in the same way.

So not only do we group liters into ten decaliters, but we also choose to split a liter into ten deciliters, and each deciliter into ten centiliters. Similarly, US dollars are broken into ten dimes and dimes into the same number of pennies.

This consistency in the way we group and subdivide allows us to extend our notation system easily. To record a sale of one hundred forty-two dollars and seventy-nine cents, for example, we could use a Hindu-style grid system like so:

| 1 | 4 | 2 ‖ 7 | 9 |

Notice the use of the double line to separate the "ones" (the dollars) from the cents. Some sort of indicator like this is necessary in order to keep track of our unit choices: Are we counting in dollars or in pennies?

Of course, just as the introduction of the zero symbol allowed us to dispense with the gridlines altogether, we can also indicate the ones column in a simpler way. This is the idea behind the so-called decimal point. Thus, we can write 142.79 and understand at a glance that we have two whole dollars and that the 7 represents seven dimes, or *tenths* of a dollar. Likewise, the 9 counts the pennies, or *hundredths*.

The upshot of this is that we can easily record and manipulate quantities that are not only unlimited in size (larger numbers merely requiring more columns on the left) but also unlimited in their precision or fineness. Each new breakdown or subdivision simply adds a new column (or "decimal place") on the right. A scientist, for example, may require as many as a dozen decimal digits to properly encode a highly accurate measurement. The Hindu-Arabic system, together with a well-chosen and consistent system of measuring units (e.g., the metric system), provides an efficient and convenient way to do this.

Let's say you are an environmental chemist, and you are testing the air quality of a certain city. A 20-gram sample of air is found to contain the following quantities of various gases (the symbol g stands for grams):

Nitrogen: 15.622 g
Oxygen: 4.2 g
Argon: 0.17 g

What is the total mass (in grams) of the remaining impurities?

The important thing is to remember what the decimal point really means: it tells us what the various decimal places signify. Here we are measuring in grams, so the column immediately to the left of the period is for counting the grams themselves (the ones), whereas the final 2 in the nitrogen measurement is counting the milligrams, or *thousandths* of a gram.

The point is that in order to combine such amounts in the usual way, we now need to align their decimal representations appropriately, so that all the ones are lined up in the ones column. Of course this just means we have to align the decimal points, and everything else will line up automatically. Thus we add:

$$15.622$$
$$4.2$$
$$0.17$$

Notice the various empty spaces, both on the left and on the right of the decimal points. If you like, you can simply fill these blanks with zeros, or even draw in the gridlines if that feels better:

1	5	6	2	2
	4	2		
		1	7	

Of course, these are minor issues of visual display and presentation, but they can be surprisingly powerful psychologically. The only thing that really matters is that we understand what we mean by our symbols and notations—that we own our language and feel comfortable and free to play and invent.

In any case, it is now a simple matter to calculate the total mass of these gasses to be 19.992 grams, meaning the remaining impurities account for 0.008 grams, or 8 milligrams, if you prefer.

Naturally, there are all sorts of conventions and protocols within the scientific community for how to produce and record physical measurements and how to indicate tolerances and known accuracy and other such issues that I'm not particularly interested in discussing. All I'm saying is that if you choose your units wisely (that is, consistently), you reap certain benefits of uniformity and avoid the unpleasantness of shillings and pence.

This is ultimately the point (and the only real value) of the metric system—it's simpler, that's all. Having all of our units (with the exception, alas, of time) grouped and subdivided into tens, in agreement with our conventional number system, makes things easier and, if anything, *less* technical and confusing.

Not that *ten* matters at all. We could just as easily imagine an alternative history, where the Tree People somehow defeated the Egyptians and ended up ruling over a vast empire, eventually establishing a septimal place-value system throughout the world. Presumably, at some point scientists and others would then adopt units that are intelligently grouped and subdivided into *sevens*.

Surprisingly, this would not look or feel all that different from our own decimal-based metric system. We could imagine that the Tree People would sooner or later hit upon the idea of a place-value abacus with marked-value entries just like our own, the only differences being that the columns would represent successive groupings by seven instead of ten and that they would thus require only seven distinct symbols (including zero if they thought of it).

In order not to drive ourselves crazy memorizing new symbols (or worse, dealing with the original Tree hieroglyphics), let's imagine the Arab traders again importing the system into Europe using the same numerals we are already used to, namely 0, 1, 2, 3, 4, 5, and 6. The advantage is that we instantly recognize and understand the meaning of these signs, the disadvantage being that our familiarity with the marked-value aspect may blind us to the substantial change in grouping size. With that small caution, let's see how such a system would work.

First of all, notice that the number seven itself (that is, one group) would then be written 10, just as we are used to writing ten. A code like 34 would mean three groups and four leftovers. You might say to yourself, "Oh, I see. That would be three groups of seven and four leftovers, so it's really twenty-five." Well, yes and no. Yes, it certainly is the same as the quantity that we are used to calling twenty-five, but no, that's not what it is *really*. That's only what it is familiarly. To the Tree People and their descendants, it's not two groups of ten and five leftovers, it's three groups of seven and four leftovers. Their sense of what an amount is and how it feels would have evolved around groups of seven, including their spoken number words. We feel that fourteen is a sloppy number full of leftovers, midway between the solid and upstanding quantities ten and twenty. Not so for the Tree People. This number is simply *two groups*, written 20 and referred to by a simple name like twotree, which means two whole groups. Counting by sevens would be almost effortless—tree, twotree, thirtree— just as we count ten, twenty, thirty.

Going further, we could even extend the system to fractional parts of a unit in the same way as before, using a "septimal point" to keep track of the meaning of the columns. Thus, the septimal digit sequence 12.3 would stand for nine full units (meters, say) along with three subdivided units (septimeters, I suppose). In other words, 12.3 would encode the number that we would normally call nine and three-sevenths. Does that make sense?

Obviously, converting from one grouping size to another is complicated and technical, and we'll have to talk about it lots more; all I'm trying to get across right now is that there is nothing at all special about the number ten, and that all the conveniences and advantages of the Hindu-Arabic system, as well as the associated metric system of units, would equally apply to any grouping size whatsoever.

> *Suppose the Hand and Banana tribes also developed*
> *place-value systems in the same way. Using*
> *modern numerals, how would you write the quantities*
> *sixteen, twenty-seven, fifty-two, and eighty-eight*
> *in the three tribal systems?*

MULTIPLICATION

At first glance, it may seem that all of these strategies and choices of how to represent and manipulate quantities are pretty much the same; that the differences between repetition and grouping, stacking and subgrouping, marked value and place value don't amount to a hill of beans (or a pile of rocks). And for most everyday counting and bookkeeping purposes, this is probably true. If you are playing a game of dominoes, it hardly matters what system you use to keep score, and in fact, the five-barred gate is probably just as good if not better than any of the fancier, more technically demanding alternatives. One might well wonder what all the fuss is about.

The truth is that for most arithmetic purposes—counting and recording quantitative information, making comparisons, keeping track of income and expenditures, additions and subtractions—all of the systems we have discussed are pretty much equivalent and have all flourished and proven their utility at one time or another (with the exception of the tribal systems that I invented out of whole cloth).

In particular, there is nothing so great about the Hindu-Arabic decimal place-value system. You could become just as fluent and proficient at using the Roman system, for example. So why do we bother? What is really to be gained by learning to use a symbolic place-value representation system? I suppose there is the convenience and portability of a pencil and paper abacus, but we could just as easily do this with the Roman system. It doesn't take any great skill, for instance, to add two Roman numerals together on paper: MMDCCLXXVII and MCCLXXXVI combine to make MMMMLXIII. You can simply count the number of occurrences of each symbol, cashing in and exchanging in your head—mimicking the maneuvers of the tabula in your imagination. This is surely no more difficult than the mental gymnastics needed for the Hindu-Arabic system.

On the face of it, it looks like the whole "place-value with marked-value entries" idea is a lot of trouble and memorization without a lot of payoff, aside from maybe making our representations a bit shorter.

There is one situation, however, that occurs rather frequently in arithmetic, where place-value representation possesses a clear and overwhelming advantage: *making copies.*

It often happens when counting and keeping track of quantities that the same number needs to be added (or subtracted) repeatedly—a dozen muffins are sold, twenty cases get shipped, we need to triple the recipe, and so forth. Essentially, this comes down to adding a number to itself several times. This is known as *multiplication* (from the Latin *multiplicatus,* "folded many times"). People say things like "six times seven" (usually abbreviated as 6 × 7), meaning that seven is to be added to itself six times; that is, we want the total of six copies of seven.

Notice that there is a subtle yet important difference in status between the two numbers here. When one writes 5 × 8, the 8 is the actual quantity we are adding—the price of a single muffin, say. The 5 is a *counter;* it indicates the number of copies we want. In a sense, the 5 is "operating" on the 8, and not the other way around.

So there is an asymmetry to multiplication: 5 × 8 does not mean the same thing as 8 × 5. Surprisingly, however, the two totals happen to be the same. That is, even though five baskets holding eight eggs each is an entirely different scenario from eight baskets of five, the total amount of eggs is nonetheless equal. My favorite way to see this is to imagine rocks laid out in rows:

$$
\begin{array}{ccccccc}
\bigcirc & \bigcirc & \bigcirc & \bigcirc & \bigcirc & \bigcirc & \bigcirc \\
\bigcirc & \bigcirc & \bigcirc & \bigcirc & \bigcirc & \bigcirc & \bigcirc \\
\bigcirc & \bigcirc & \bigcirc & \bigcirc & \bigcirc & \bigcirc & \bigcirc \\
\bigcirc & \bigcirc & \bigcirc & \bigcirc & \bigcirc & \bigcirc & \bigcirc \\
\bigcirc & \bigcirc & \bigcirc & \bigcirc & \bigcirc & \bigcirc & \bigcirc \\
\end{array}
$$

Here we have a simple visual representation of five copies of eight. Each row contains eight rocks, and there are five such rows. So 5 × 8 is five rows of eight. On the other hand (and here's the clever part), we can also view this layout of rocks as consisting of *columns*—namely, eight columns of five rocks each. Another way to say it is that by turning our heads sideways, the five rows of eight become eight rows of five. Thus 5 × 8 equals 8 × 5 because both are counting the same rectangular array of rocks. So multiplication turns out to be symmetrical after all. This is not only unexpected and pretty, but it also turns out to be quite useful and important.

For example, if I were called upon to add up seventeen copies of two (that is, 17 × 2), rather than heaving a deep sigh and proceeding to add twos together until the cows come home, I could cleverly make use of symmetry to reimagine this as 2 × 17 (i.e., seventeen doubled), which is a much simpler computation. This picture of multiplication as a rectangle of rocks also clearly pertains to the measurement of areas such as a rectangular plot of ground, or a wall that needs to be plastered and painted. Usually, such areas are sectioned off into a grid, and then the grid squares can be counted using multiplication.

Let's suppose you are tiling the bathroom floor, which measures six by eight feet. Linoleum tile comes in one-foot squares, fifty to a box. Will one box be enough to get the job done?

Here we can imagine the floor sectioned into one-foot grid squares:

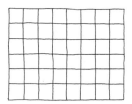

We could, of course, simply count these squares one by one, and in fact that wouldn't really be all that bad; there aren't that many of them, and it would only take a few seconds. On the other hand, it is valuable to view this as a multiplication (namely, 6 × 8), not only for the insight and the intellectual pleasure it may give us to think of it this way but also because as the numbers get larger (e.g., 173 × 62), counting the squares individually becomes increasingly obnoxious and time consuming.

So the question is whether 6 × 8 is smaller or larger than 50. Here is the fundamental arithmetic issue again: comparison. On the one hand, we have a number (6 × 8) simply expressed and nicely organized into groups of eight, which we want to compare to another number, fifty, which is being held linguistically as five groups of ten. So it's just like the Hand People trading with the Banana People. We have two different grouping sizes, and we want to translate between them so we can compare.

One way to proceed would be to stick with ten as our grouping size and try to rewrite six groups of eight in terms of tens, then see how it compares to five groups. This is what most people mean when they say they are multiplying: converting a quantity conveniently expressed in one grouping size into piles of ten and leftovers. Let's try to do it visually with rocks, sliding them around to make nice rows of ten:

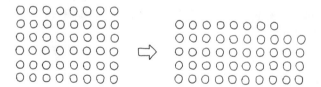

Taking rocks away from the first row, I can add two rocks to each of the four bottom rows, making nice rows of ten, and ending up with four complete rows and eight leftovers. Thus we discover that six groups of eight is the same as four groups of ten and eight leftovers. In other words, 6 × 8 = 48.

In this way, we can think of the number 48 as having been "produced" from 6 and 8 via multiplication. People say that 48 is the *product* of 6 and 8.

Alternatively, we could simply add eight to itself six times, cashing in and exchanging when necessary. One amusing way to proceed would be to start by doubling 8 to get 16, then adding on another 8 to get 24. This counts three copies of eight, so to get all six copies all I need to do is double this number to get 48. This is what I mean when I say that arithmetic is the art of rearranging quantities—noticing and taking advantage of specific features of your counting problem to make your life easier and to have a bit of amusement while you're at it.

Either way, we now have our number expressed in groups of ten, so it is easy to do the comparison and see that, in fact, one box of fifty tiles will indeed do the job. (In real life, of course, things are never that simple. You'll have to cut a few tiles to get around the bathtub and the toilet, and you'll probably also ruin a few by accident, etc.)

Probably the most frequently occurring instance of multiplication is *doubling*—adding a number to itself. Of course this is just a special case of adding two numbers together. Because it occurs so often, though, people who do a lot of arithmetic tend to become very accustomed to the way doubling behaves and get quite familiar with the doubles of small numbers in particular. Thus, one comes to learn and remember such things as "six doubled is twelve" and "two fours make eight" and so on. This would be true regardless of what representation system you happen to use; a Banana tribesman would also know from experience that *na–na* doubled is *ba*, and twice *na–na–na* is *ba–na–na*.

In marked-value systems, doubling is almost literally just that: every symbol, counting coin, or calculus stone is simply repeated. Exchanges can then be performed if necessary.

> *Calculate the doubles of the following numbers,*
> *packed up and written in their respective languages:*
> *ba-ba-na-na,* ? ⩑⩑ ⦚⦚⦚, MCCCLXXVI, 538.

The Egyptians were particularly fond of doubling and used it in very clever ways to calculate larger multiples. To triple a number, for instance, we can double it and then add on the original number. To quadruple a number (that is, to make four copies), all we need to do is double it and then double again. To multiply by six, we can double our number, then redouble it and add these two together.

What would be a good way to use doubling
to multiply by five? How about by twelve?

Multiplication by small numbers like three or five is particularly easy to do using a simple abacus system like Piles of Rocks, counting coins, or the Roman tabula. We simply lay our number out as many times as we need, then cash in and we're done. The soroban is somewhat less convenient, as is the Hindu-Arabic pencil-and-paper abacus. In particular, both of these require a considerable amount of memorization and mental gymnastics. When I have occasion to double something like 427, for instance, I tend to think of it like this: "OK, I have four hundreds, two tens, and seven leftovers. Doubling everything gives me eight hundreds, four tens, and fourteen. So, eight hundred forty-fourteen, which is eight hundred fifty-four, or 854." Similarly, tripling in my head would give me twelve hundred, sixty, twenty-one; in other words, 1281.

Of course, with a purely symbolic abacus like this, it pays to know your small multiples—like how I happen to remember that three sevens is twenty-one. This can be a bit of a chore, especially with a large grouping size like ten. I do *not* recommend that you intentionally set out to memorize such things. It's far better to simply play around a lot with numbers, get used to using your own clever strategies and observations, get occasionally frustrated with your inability to recall, and gradually these facts will become familiar and friendly and may even carry with them pleasant reminders of a problem you enjoyed or a moment of cleverness on your part.

One thing that is often done is to make a chart (the infamous "times tables") that lays out the various small multiples in tabular form. This can be a handy thing to have around if you are going to be doing a lot of multiplying, and it can also be an easy way to notice many amusing patterns. The danger is that in attempting to memorize such a mass of data you will get bored and frustrated and turned off to arithmetic. So my advice would be to make such a chart if you wish to consult it but don't stress out about it or put too much importance on it. After all, we live in the so-called Information Age. If I can't remember the capital of Turkmenistan, I can always look it up; the same goes for 7 × 8 (which is one that has always been a problem for me).

Oh, right. It's 56. While we're on the subject, let me just emphasize a couple of things. First of all, let's understand that 7 × 8 is not a question and 56 is not an answer. Seven times eight is a *number*, and it is capable of being represented in a great many ways. At the moment it is held as seven groups of eight, and a user of an octal system would be quite pleased and would not feel the need to "do" anything to it. When we ask, "What is seven times eight?" what we are really asking is, "How can seven groups of eight be rearranged into groups of ten for ease of comparison with other numbers similarly grouped?" Numbers couldn't care less what grouping size you happen to use and neither do mathematicians. Numbers are what they are, and they do what they do; your desire to *compare* is the issue, and your culturally determined choice of representation system is quite secondary.

Another thing to notice is that when we say something like "seven times ten is seventy," there is an amusing content-free circularity to such a statement. After all, the word *seventy* is simply an abbreviated form of seven tens.

Languages always reflect the culturally agreed-upon grouping sizes, and the names for these numbers will then become the familiar benchmarks for size and quantity. To say that seventy is the answer to the question "what is seven times ten?" is particularly ludicrous. It's like looking up a word in the dictionary only to find it defined in terms of itself. What seven times ten *is* is seven groups of ten. In a decimal culture one would then already be quite satisfied—the number is already expressed in a way convenient for comparison. The Banana People would be less happy and would want to start arranging this quantity into bunches of four.

How would the Tree People feel about seventy?

One frequent situation where multiplication comes into play is the sale and purchase of several items of equal value. For instance, here is a typical receipt from a nineteenth-century silversmith:

Bill of Sale

Qty	Description	£	s.	d.
6	silver spoons, 4s. 5d. ea.	1	6	6
3	snuffboxes, 1£ ea.	3		
2	candlesticks, 10s. 6d. ea.	1	1	
	Total	5	7	6

Here the silversmith has cleverly done the separate multiplications mentally, which I imagine one would eventually get quite used to doing.

Is the silversmith's total correct?

Let's look at a more ancient version using the Egyptian system: one hundred bushels of wheat remain in the storehouse. You, the miller, have recently received several orders for grain. The local baker needs seven large baskets of wheat, the temple has ordered eight small baskets, and the pharaoh's tax collectors are to be given five large baskets as well. The small baskets hold three bushels, and the large ones hold six bushels. Do you have enough?

Here the arithmetic problem is to determine whether the total of seven copies of six, eight copies of three, and five copies of six is safely below one hundred.

There are, of course, many ways to proceed from here, the most likely being that you, as an experienced tradesman, would simply know what these amounts come to (in terms of groups of ten) and can tot them up in your head. But let's suppose that you are only an apprentice miller and are relatively unskilled at mental calculation. Then you could simply get out your enormous bag of marked-value counting coins and proceed to lay these amounts out on the counter. Gathering together groups of ten and making the necessary exchanges would certainly get the job done, albeit in the most boring and rote mechanical way. Instead, let's try to be a bit more creative. Being Egyptians, we'll begin by doing a little doubling:

> one copy of six: |||
>
> two copies of six: ||| ||| , or ∩ ||
>
> four copies of six: ∩ || ∩ || , or ∩∩ ||||

Now we can get seven copies of six by totaling these amounts (since seven is one plus two plus four). This makes a total of

$$||| \; ∩ \; || \; ∩∩ \; |||| , \text{ or } ∩∩ \; ||$$

Similarly, we see the five copies of six as being four copies plus one more copy, that is

$$∩∩ \; |||| \; ||| , \text{ or } ∩∩∩$$

Now, the eight copies of three we can cleverly view as four copies of six (each pair of threes makes a six), so we already have that as ∩∩ ||||. Thus, our grand total is

$$∩∩ \; || \; ∩∩∩ \; ∩∩ \; |||| , \text{ or } \widehat{∩∩∩} \; ||| ,$$

and we are just a few bushels under ⌐. So there actually is enough in the storehouse, and we needn't panic.

*How would you do this same calculation
using the Roman tabula?*

In a symbolic place-value setting, such as the Hindu-Arabic system, the situation is pretty much the same, except we have no counting coins or calculi to jiggle around. We must manipulate symbols with our minds, instead of objects with our hands.

Just as with addition and subtraction, this presents us with a bit of a challenge. If we wish to become truly fluent users of such a system, we will need to figure out how the symbols transform (e.g., $6 + 7 = 13$, $4 \times 3 = 12$), and commit such patterns to memory.

Again, my advice is to work them out from scratch each time, using your fingers, rocks, tally marks, or whatever. Yes, it will be annoying. But the annoyance is what will make them stick in your mind. It doesn't matter how long it takes (nor does it matter if it ever happens). If you like to count and you play around enough, eventually all of these things will become quite familiar.

Suppose you're at your neighbor's yard sale and you find three shirts for seven dollars apiece. All you have is a twenty. Can you get all three shirts?

The question is whether 3×7 is safely below 20. Laying out three rows of seven rocks, we can move a few from one row to build two nice rows of ten:

Thus we discover that 3×7 is in fact 21. So we'll either have to leave one of the shirts, borrow a dollar, or do some haggling. (Since it's a yard sale, they'll probably just let you have them.)

*Calculate the doubles and triples of each number
from one to nine by rearranging rows of rocks.*

So far, the differences among our various representation systems with respect to multiplication have not been all that great. Whether you are using a repetition system (like tally marks), a marked-value system (like the Egyptians), or a place-value system (such as the Roman tabula or the Hindu-Arabic system), as an experienced arithmetician you will simply come to know all of the small multiples by memory, and can do most of the exchanges and so forth in your head. A moderately talented Egyptian, Roman, Japanese, or Indian scribe would easily calculate seven copies of one hundred sixty-three in more or less the following way:

"Let's see. I've got seven hundreds, plus another seven copies of sixty—that's seven copies of six tens. Since seven sixes is forty-two, that makes forty-two tens, which means four hundreds and two tens. So really I've got eleven hundreds, two leftover tens, and then don't forget the three sevens, which is twenty-one. So eleven hundred forty-one." He would then proceed to write it down in whatever representation system he was using. Only the most elaborate (or delicate) calculations would require getting out a soroban or tabula.

In many ways, the fun really begins when you try to do your calculations mentally like this. Of course, you will mess up a lot and get delightfully lost and confused. It's a tricky and challenging exercise, like playing chess blindfolded. There are a bunch of simple rules and patterns, and you have to keep it all straight. But the stakes are quite low, and it can be pretty amusing at times. Also, a little familiarity with numbers and their behavior can lead to larger questions and curiosity about number patterns in general. Arithmetic can be a gateway drug for mathematics. So play around and make all kinds of ridiculous and nonsensical blunders—I certainly do!

The multiples of five have a particularly simple
pattern in the Hindu-Arabic system.
What is the pattern and why does it work?

Being so close to the grouping size, the multiples
of nine and eleven also make nice patterns.
Can you figure out what they are?

I want to point out something about that last calculation that turns out to be pretty useful and important. When the time came to make seven copies of sixty, our experienced scribe thought of it in an interesting way. Sixty literally means (and is an abbreviation for) six tens. So we're talking about seven copies of six copies of ten. This kind of thing happens quite frequently, in fact, where we have some amount and we multiply it by some number, and then that entire amount gets copied as well.

One way that such "triple products" occur naturally is in the calculation of three-dimensional volume. Just as we were able to view the product of two numbers as being a rectangle of rocks, we can think of something like 4 × 5 × 6 as being a *box* of rocks, like so:

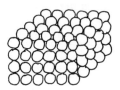

There are several ways we can imagine counting such a block. Thinking of it as a stack of horizontal plates, we can see that each layer is a 5 × 6 rectangle. There are four of them, so we have 4 × (5 × 6). On the other hand, we can also view the same box as comprising six vertical sheets, each of which is a 4 × 5 rectangle. This way, we can see the total as 6 × (4 × 5). Still another way is to imagine it as a stack of five 4 × 6 rectangles coming up out of the page, giving us 5 × (4 × 6). The point being that not only is multiplication pairwise symmetric (e.g., 4 × 5 = 5 × 4), but it is also symmetric in this larger sense—that when several numbers are multiplied together, it doesn't matter what order we do things in. This not only spares us from having to worry about parentheses, it also allows us to be creative and to choose an ordering that is clever and efficient.

In the scribe example, this meant thinking of $7 \times (6 \times 10)$ (i.e., seven copies of sixty) as $(7 \times 6) \times 10$—in other words, as a number of groups (namely, 7×6 of them). This makes forty-two groups. The number of groups of ten is itself four groups and two leftovers. So in total we have four groups of groups and two remaining groups, if you follow me. Since the only thing the word *hundred* ever meant was a group of groups, we see that the total is four hundred twenty. That's certainly a lot easier than adding sixty to itself seven times, if you ask me.

Of course, being a mathematician, I'm *always* looking for ways to avoid tedious mundane labor. Especially if it means I get to do a bunch of interesting abstract thinking. That's kind of the whole math thing: working hard to find ways to get out of working hard. If you are really, really lazy, and also happen to be really, really clever, then math just might be the life for you (assuming you also have no interest in wealth, fame, or popularity).

While we're on the subject of numbers in the abstract, especially their behavior with respect to multiplication, let me just mention an extremely important, though patently obvious fact. When we say that two plus three is five, what we mean is that two of *anything*, together with three more of those same things, whatever they may be, makes a total of five of them. Two cows added to three cows makes five cows, two years added on to a prison sentence of three years makes five years in the slammer, and so on.

In particular, this means that two dozen and three dozen make five dozen; two copies of thirty-seven together with three more thirty-sevens makes five thirty-sevens. Turning this around, we can say that if a number is multiplied by a sum of two numbers, it is the same as if we multiplied it by each of the numbers separately and then added the results:

$$(2 + 3) \times 37 = (2 \times 37) + (3 \times 37).$$

At first blush, this may seem a rather trivial and pedantic observation. But it turns out to be an extremely powerful arithmetic tool. It means we have the freedom to *break up* our multiplications into smaller pieces as we see fit, and this freedom can be used cleverly and creatively to make calculating simpler and easier.

Before we look at such clever maneuvers, let's first understand this breaking up principle in terms of rocks and rectangles.

```
O O | O O O O O
O O | O O O O O
O O | O O O O O
O O | O O O O O
```

Here we have a four by seven array of rocks, sliced into two smaller rectangles. The seven columns of four have been broken into two parts: the one on the left being two columns of four and the one on the right five columns of four. So we see 7 × 4 broken into 2 × 4 and 5 × 4.

Alternatively, we could think of this subdivision as saying that four rows of seven is the same as four rows of two together with four rows of five. Symbolically,

$$4 \times 7 = (4 \times 2) + (4 \times 5).$$

That is, when making copies of a sum, we can simply multiply each piece separately and then add them. It's like when you order two pizzas and three root beers and then you decide to double the order: the pizzas and the root beers are both doubled separately.

Of course the same is true no matter how many parts you break a number into.

Here we have 5 × 12 viewed as 5 × (3 + 4 + 5), and naturally this shatters into rectangles as (5 × 3) + (5 × 4) + (5 × 5). So the principle is completely general.

In fact, we can even go a step further and imagine both rows *and* columns broken into pieces to form a tiling of our rectangle by smaller rectangles of various sizes.

```
○ ○ ○ │ ○ ○ ○ ○ │ ○ ○ ○ ○ ○
○ ○ ○ │ ○ ○ ○ ○ │ ○ ○ ○ ○ ○
──────┼─────────┼──────────
○ ○ ○ │ ○ ○ ○ ○ │ ○ ○ ○ ○ ○
○ ○ ○ │ ○ ○ ○ ○ │ ○ ○ ○ ○ ○
○ ○ ○ │ ○ ○ ○ ○ │ ○ ○ ○ ○ ○
```

Here is our 5 × 12 rectangle again, only this time we are thinking of it as being the product of (2 + 3) and (3 + 4 + 5). What this tiling shows is that

$$(2 + 3) \times (3 + 4 + 5) = (2 \times 3) + (2 \times 4) + (2 \times 5) +$$
$$(3 \times 3) + (3 \times 4) + (3 \times 5).$$

What's happening here is that each number in the first sum is being multiplied separately by every number in the second sum. This is really the whole thing about grids and rectangles— they chop up nicely. The pattern that we've discovered here is probably the single most important property in all of arithmetic: *the product of two sums is equal to the sum of the individual products.* Of course, we have to know which products to add together— namely, all the possible choices of one number from each sum.

What happen if we multiply three sums together?

So how does this chopping up idea help us multiply more easily? Of course, having knowledge of the way something behaves is always going to be helpful, if for no other reason than to increase familiarity and understanding. If you happen to have already figured out 5 × 17 and 3 × 17, then to get 8 × 17 you need merely add them together. So that's nice to know, even if it is pretty obvious.

The real power of this method comes when we combine it with a place-value representation system that uses a consistent grouping size, such as the Hindu-Arabic system. The reason is that now there is at least one number that is always easy to multiply by: *the grouping size itself.* This is such an important discovery that it pays to look at it a bit more generally.

Let's suppose we have a marked-value system like the Egyptians. Here we have a consistent grouping size (ten) and marked-value symbols I, ∩, ੧, 𐦅 for each stage of collection. What happens if we have a number represented in this language, say ੧੧ ∩∩ III, and we multiply it by the grouping size? We will then have ten copies of every symbol. I suppose we could simply write down all these copies and then start cashing them in, but there is a much simpler (and way more efficient) procedure: simply *replace* all the I's by ∩'s! And of course each ∩ becomes a ੧, and each ੧ is transformed into a 𐦅. What amazing Egyptian alchemy! (The word *alchemy*, by the way, comes from Khem, the ancient name for Egypt.)

The point being that since ten is our grouping size, *of course* ten ones make a group. Ten of *any* symbol makes the next higher symbol. This is the benefit of having the grouping sizes be uniform and consistent. Multiplication by the grouping size then becomes a simple symbol-manipulation game. Similarly, if you had some pennies and dimes in your pocket, multiplying your money by ten would be tantamount to somehow magically transforming each penny into a dime and each dime into a dollar. (Notice that no such simple transformation would work for shillings and pence.)

The same holds for the Roman system. Multiplication by ten would have the effect of turning Is into Xs, Vs into Ls, Xs into Cs, and so on. Things are even more elegant on the tabula: multiplication by ten simply moves each calculus stone up to the next corresponding line. All you have to do is shift all the calculi on the I-line to the X-line, the V-line stones to the L-line, and so on. That's pretty convenient.

What happens when the Banana People
multiply by their grouping size?

Of course, it's really with the Hindu-Arabic system that things become almost effortless. Here, since each digit symbol sits in a column whose position controls its value, the effect of multiplication by ten is simply to *shift* each symbol one place to the left.

For example, to multiply a number like 137 by ten, we just shove each digit over one place:

	1	3	7	
1	3	7		

Now the 7 symbol, instead of counting ones, is counting tens—as if each leftover were "upgraded" to a group. What could be easier than that?

By the way, a lot of people are under the mistaken impression that to multiply a number by ten, you simply add a zero on to the end of its digit sequence: 137 becomes 1370. In a sense this is true—at least visually. But I think it's important to understand that we're not really adding a zero on the end; it's that the whole sequence is being *shifted*, so that each symbol is counting a collection ten times larger than before.

I could come up with several arguments in favor of the Hindu-Arabic system over its marked-value cousins—lightweight portability, low cost, and so forth—but the truth is that it is this place-shifting business that is the real advantage. No transmutation of coins or symbols, no sliding rocks around from one line to another. It's just symbols lightly dancing on the page, leaping from one column to the next. That's the ultimate reason why the Roman system had to make way and why pretty much the whole world now uses the Hindu-Arabic system.

In fact, this place-shifting idea gives us a whole new way to view numbers such as 20, 500, and 3000. I like to think of numbers like these as being single-digit numbers, just like 2, 5, and 3. The number 500, for instance, is really just 5 (hundreds). I'm certainly not saying that five lemons is the same as five hundred lemons. I'm just noticing that both quantities are counting five of something, and those somethings are simply our various places—the slots or columns in our Hindu-Arabic representation scheme.

So let's start thinking of 500 as being 5 double-shifted. Then 20 is 2 shifted, and 3000 is the same as 3 with a triple shift. Each shift corresponds to a multiplication by ten, so another way to say this is that $3000 = 3 \times 10 \times 10 \times 10$. The only thing these multiplications by ten do is cause some shifting, so in some sense the real content is the 3—what scientists and engineers like to call the *significant digit*.

The reason this is a useful viewpoint is that when two such single-digit numbers are multiplied together, something very simple and pretty happens. Let's take for example 30×200. Of course you are free to write down thirty copies of two hundred and add them all up in whatever way suits you. That is, after all, what multiplication means—repeated addition. But who wants to do that? Not only would it be tedious and boring but, as we've talked about before, the tedium would undoubtedly lead to sloppy, error-prone miscalculation. The whole point of arithmetic is to be cleverer than that so that computation is faster and easier and more fun.

All right, so let's think about 30×200 in a better way—and I don't mean two hundred copies of thirty! Instead, let's view it as

$$(3 \times 10\ ) \times (2 \times 10 \times 10).$$

Now we can take advantage of all the symmetries we've discovered to rewrite this five-term product as

$$(3 \times 2) \times (10 \times 10 \times 10) = 6000.$$

Notice that the two digits at the head of our original numbers (namely, 2 and 3) are getting multiplied, and that the shifts are being compounded into one big total shift. So 3 with one shift, multiplied by 2 with two shifts, comes out to be 6 with three shifts. So it's particularly easy to multiply these sorts of single-digit numbers: just multiply the digits and add the shifts. Does that make complete sense?

Calculate the products: 20 × 40, 30 × 500, 800 × 5000.

What a fantastic time and energy saver this is! Of course, you may be saying to yourself, "Yeah, that's great and all, but how often am I going to be in the luxurious position of multiplying such simple numbers together? How does this help with some nightmare like 37 × 168?"

What we would like, if possible, is a simple procedure whereby we can take the digit sequences of the two numbers alone (without the need for any rocks or coins or beads) and produce—in a reasonably quick and efficient (and hopefully accurate) way—the digit sequence of their product. It turns out that for the Hindu-Arabic system, this is, in fact, quite possible.

The idea is to combine the shifting business with the breakdown strategy we talked about earlier. Let's start with some relatively small numbers, say 12 × 27. Imagining this as a twelve by twenty-seven rectangle, we can break it down into smaller rectangles in many ways. The simplest and most natural breakdown (at least from the ten-centric point of view) is to chop each number into its single-digit pieces, like so:

	20	7
10	10 × 20	10 × 7
2	2 × 20	2 × 7

Notice that each subrectangle becomes a product of single-digit numbers. These are easy to calculate by shifting:

$10 \times 20 = 200$, $10 \times 7 = 70$, $2 \times 20 = 40$, and $2 \times 7 = 14$. (Here's where knowing your small multiples comes in handy.) So, putting the pieces together, we get

$$12 \times 27 = 200 + 70 + 40 + 14 = 324.$$

Yes, we have to do some adding here at the end, and yes, there may be some nasty carrying and exchanging to do, but you can't get something for nothing. At least we didn't have to add a dozen twenty-sevens together!

But of course the power of this method lies in its complete generality—we need no additional ideas in order to multiply *any* two numbers together, no matter how large. The only thing that will change is the number of pieces the rectangle shatters into.

Let's imagine that you run a pickle factory. Each barrel of pickles weighs 144 pounds fully packed. You want to ship 56 barrels, but the delivery truck can carry a maximum of 8000 pounds. Can you do it in one shipment?

Here the question is whether 56×144 is safely less than 8000. Of course we could just get out a bunch of rocks or do a ton of repeated addition, but the point of our new method is to take advantage of our place-value encoding so that we can perform such calculations purely symbolically, without any rocks or beads, and also quickly and efficiently, minimizing redundant labor.

In this case, our (imaginary) rectangle would break down like this:

	100	40	4
50	5000	2000	200
6	600	240	24

Here I've filled each subrectangle with the corresponding number of imaginary rocks. Thus, the 50×40 rectangle receives the entry 2000, since $5 \times 4 = 20$ and we have another two shifts. This means the total weight of all the barrels together is

$$5000 + 2000 + 200 + 600 + 240 + 24 = 8064.$$

So we are sixty-four pounds overweight. You may decide to risk it, or to leave a barrel or two off the truck, but at least you know where you stand.

Which is larger, 381×44 *or* 598×28 *?*

Of course, once we have a purely symbolic algorithm like this, especially if it makes good sense and we understand why it works, then naturally we get lazy and impatient and we want to streamline the procedure to make it even easier. One possible first step is to eliminate the imaginary rectangle altogether. It certainly is a useful device for explaining to ourselves what we are doing and provides a clear visual image of how we are organizing our counting, but in fact it is in no way essential to the actual performance of the calculation. Once we fully understand the breakdown and place-shifting principles, we no longer need a pictorial representation.

For example, to multiply 173 by 254, we can simply run through the various possible single-digit products (in whatever order we wish), making sure we include every possibility and account for all the shifts:

$100 \times 200 = 20000$	$100 \times 50 = 5000$	$100 \times 4 = 400$
$70 \times 200 = 14000$	$70 \times 50 = 3500$	$70 \times 4 = 280$
$3 \times 200 = 600$	$3 \times 50 = 150$	$3 \times 4 = 12$

These can then be totaled as usual. One popular way to organize this kind of work on the page is to write down the two numbers you wish to multiply, one below the other (properly aligned, of course), and simply enter the single-digit products below:

```
 173
 254
  12    4 × 3, no shifts
 280    4 × 7, one shift
 400    4 × 1, two shifts
 150    5 × 3, one shift
3500    5 × 7, two shifts
5000    5 × 1, three shifts
 600    2 × 3, two shifts
14000   2 × 7, three shifts
20000   2 × 1, four shifts
43942
```

This way, the various pieces are all lined up in the correct columns and ready to be totaled. Of course, if this seems too abstract and you prefer to draw imaginary rectangles of rocks, then by all means do that. Eventually you will come to see the pattern to how all the parts are generated and assembled, and you will no longer require its services. This is what we humans always do once we've mastered something—look for shortcuts.

In fact, many people choose (or have chosen for them) an even more streamlined version, wherein the pieces are subtotaled as they are produced. This requires a bit more mental labor and memory, but it's quite a bit faster and takes up considerably less space. Apparently, this has become the standard algorithm taught in elementary schools. Let's see if I can explain it.

To begin with, let's imagine you want to multiply a largish number, say 1783, by a one-digit number like 4. With our current scheme, this computation would appear like so:

```
1783
   4
  12
  32
  28
   4
7132
```

Notice that I have left off writing all the zeros and have used the column alignment itself to keep track of all the shifting— less writing but more care required.

One thing to observe is how the various shiftings make it that there are at most two digit symbols in each column. This is because the product of two digits can be at most $9 \times 9 = 81$, which is a two-digit number. Also, each line is shifted at least one place over from the previous. So never will we have three digits in any one column.

This means you can attempt to total these numbers at the same time you are producing them, provided you can hold each digit in your mind long enough to produce each successive single-digit product. Yet another good reason to commit the times tables to memory—so that you can produce the pieces fast enough not to forget them.

Using our sewing needle metaphor, this calculation would proceed as follows:

$$
\begin{array}{r}
1783 \\
\underline{4} \\
2
\end{array}
$$

Starting with the ones digits on the right, we have $4 \times 3 = 12$. All future pieces will be shifted, so the 2 really is the ones digit of the total. We may as well write it down in its correct place. The 1 (which is really a ten) we store in our minds for a second—it's on the needle, so to speak.

Now, for the next stitch, we have 4×8, although the 8 is shifted, so it's really eight tens. Thus we have our next piece, which is 32 shifted (I supposedly know from memory that $4 \times 8 = 32$). If you look back at our previous version of this computation, you can see these first two pieces, 12 and a shifted 32. Notice how the 1 and the 2 line up in the tens column. This is because the 1 is counting the tens in the number 12, and the 2 is recording the tens in 320, the shifted version of 32. The point being that we need to add them together. That is, the ones digit of 32 gets combined with the carried 1 from 12, which is currently residing on the needle.

This all feels a bit awkward to describe, let alone read, so I hope you can follow me. In some ways I'd almost prefer to let you figure it out for yourself. But we've come this far, so let's see how it all plays out (and also see what a mess I make trying to explain it).

We've gotten to this point:

$$
\begin{array}{r}
1783 \\
4 \\
\hline
32
\end{array}
$$

Here is the 3 (coming from 32 + 1 = 33 tens) and there is also a 3 on the needle. The next stitch (which gives us the hundreds) would be 4 × 7 = 28, plus the 3 on the needle makes 31, so we write down the 1 and store the 3 on the needle:

$$
\begin{array}{r}
1783 \\
4 \\
\hline
132
\end{array}
$$

Finally, the thousands digit is obtained from 4 × 1 = 4, plus the three on the needle makes 7:

$$
\begin{array}{r}
1783 \\
4 \\
\hline
7132
\end{array}
$$

Frankly, if you ask me, I'm not sure all this needlework is really worth it. I actually prefer our previous method. It's a little more writing but far less confusing and mentally taxing. Still, if you like such things, the fancy version is there as an option.

The full-blown implementation of this algorithm would look like this:

$$
\begin{array}{r}
1783 \\
\underline{274} \\
7132 \\
12481 \\
\underline{3566} \\
488542
\end{array}
$$

Here each line is produced by needlework in the way we just saw, remembering to shift each successive line one place to the left to account for the place values. For instance, the 7 in 274 is really a shifted 7, so the entire product 7 × 274 = 12481 must also be shifted. In any case, one still has to total all the pieces; there are just fewer of them.

How do our two methods compare when
the numbers get extremely large?

One thing I want to make sure is clear—because it really is the whole point here: these symbolic multiplication algorithms are nothing more than a natural consequence of the rectangle breakdown idea, together with the happy observation that in a consistent place-value system multiplication by the grouping size is equivalent to place shifting.

In particular, these methods have nothing whatever to do with the number ten. They will work just as well for any grouping size whatever. Rectangles chop up into smaller rectangles regardless of how you choose to represent quantities. This is a mathematical feature, not a cultural one.

To illustrate, let's imagine that the Banana People have independently hit upon the place-value idea and have replaced their repetition-based, marked-value system with Hindu-style columns. We'll suppose further that they also came up with the zero symbol idea to obviate the need for column lines.

Of course, they would then also need marked-value symbols for their leftovers, namely one, two, and three. The specific choice of symbols doesn't really matter, but they should at least be easy to write down and recognize. Let's imagine they settled on these:

$$0 \qquad I \qquad \Lambda \qquad \Delta$$

| zero | one | two | three |

(I like how each digit symbol is made of the corresponding number of strokes.) Thus we obtain a "quaternary", or base-four place-value representation system. The number that we call twenty-seven, which would formerly have been rendered in Old Banana as ⊞□□\\\, would now be written more succinctly as I$\Lambda\Delta$.

In particular, the Banana grouping size (four) would be represented by I0, just as our conventional grouping size (ten) is written 10, the point being that a group (in whatever language) is always one group and no leftovers. Notice also that the shifting thing still applies. To multiply by four is the same as converting \'s into □'s, □'s into ⊞'s. This means that in the place-value version, we are simply shifting one place to the right, just as before.

So the only really substantive change to our addition, subtraction, and multiplication procedures is that the single-digit sums and products have different representations. For instance, instead of memorizing $3 \times 3 = 9$, a Banana tribesman would want to remember $\Delta \times \Delta = \Lambda I$. Notice that now the addition and multiplication tables are much smaller. In fact, here they are in complete totality:

+	I	Λ	Δ
I	Λ	Δ	I0
Λ	Δ	I0	II
Δ	I0	II	IΛ

×	I	Λ	Δ
I	I	Λ	Δ
Λ	Λ	I0	IΛ
Δ	Δ	IΛ	ΛI

That's not a lot of memorizing to do. So a smaller grouping size is great for cutting down the mental workload. On the other hand, it does mean our representations will get somewhat longer. For instance, our number 1783 from before would now be expressed by the Banana digit sequence IΛΔΔIΔ. Longer sequences mean more shifting and hence more steps in the algorithms. So you save time memorizing, but you pay later in calculating. Of course, as we will see, all of these step-by-step arithmetic procedures have been fully mechanized, and thus so-called pencil-and-paper arithmetic is now a rather obsolete skill. So get as good at it as you wish to. The important thing is to understand the ideas—at least if you want to be an arithmetician of wit and intelligence.

The multiples of Λ, Δ, *and* II *have particularly nice patterns in the Banana system. What are these patterns? How do they compare with the multiples of* 5, 9, *and* 11 *in the Hindu-Arabic system?*

Amusingly, our previous calculation of 4 × 1783, which was somewhat involved using the Hindu-Arabic system, is now a complete triviality in the Banana system. After all, four is our grouping size. So we would just take our number IΛΔΔIΔ and shift it one place to get IΛΔΔIΔ0. This could then be easily compared to other quantities similarly expressed. (And without some sort of eventual comparison, there would be no point in doing the computation in the first place.)

Let's try a more complicated example, say IΛΛ × ΛΔ (this would be 26 × 11 in decimal). Organizing our single-digit products and shifts just as before, and either consulting or memorizing our small quaternary times tables, we get:

$$
\begin{array}{r}
I\Lambda\Lambda \\
\underline{\Lambda\,\Delta} \\
I\,\Lambda \\
I\Lambda0 \\
\Delta00 \\
I00 \\
I\,000 \\
\underline{\Lambda000} \\
I0\,I\Delta\Lambda
\end{array}
$$

113

And of course, if you wished, you could streamline this further with needles and so on. The point here is that place value and multiplication are friends; the particular grouping size is irrelevant to their friendship. This is why all of the earlier marked-value systems have gone by the wayside.

While we're on the subject, an extreme case of small grouping size would be a *binary* system, with a grouping size of two. There would be only two symbols needed, **0** and **1**. The only thing to remember is that **1 + 1 = 10**; there are no pesky times tables at all. Of course, the price we pay is having extremely long digit sequences. For example, the number one hundred would be encoded as **1100100**. Now *doubling* is accomplished by place shifting. Everything works fine; multiplication becomes a babyishly simple but quite lengthy procedure.

> *Calculate* **110 × 111**. *What do these numbers correspond to in the Hindu-Arabic system?*

Incidentally, the whole place-shifting idea still works perfectly well in the case of subdivided units, provided the subdividing is consistent with the grouping size. For instance, 129 pennies when multiplied by ten would be 1290 pennies (one shift to the left), but if we preferred to count in dollars, we could just as easily say that 1.29 becomes 12.9. Again, pennies become dimes, and dimes turn into dollars. The consistency of the representation system (each column being worth exactly ten times as much as the previous) allows for a simple visual interpretation of multiplication by ten—namely, place shifting.

In particular, this gives us an alternative way of viewing numbers like 1.29, as being 129 with two *antishifts*. This means that for numbers represented in such a form, division by ten is just as simple as multiplication—we just shift to the right instead of the left.

Some people are in the habit of thinking that multiplication or division by ten moves the decimal point one space over, as with 10 × 1.29 = 12.9. This is certainly apparent visually. But really what is happening is that the decimal point is staying put, and the *digits* are moving, shifting over so that their value changes. It's obviously a minor point, but I think it's worth keeping in mind.

What this means, of course, is that we needn't make any changes to our methods in order to incorporate these sorts of representations. The crucial thing is to keep all the alignment information intact. As we've seen, to add numbers of this form together, we need to make sure the decimal points are aligned, else we will be adding apples and oranges (or at any rate, ones and hundreds).

In a way, multiplication is a bit simpler in this regard, because the various place shiftings can be dealt with independently. For example, if we wanted to calculate 12 × 3.85, we could just as easily view this as being 12 × 385 with two antishifts that can be performed later if we desire. This is yet another advantage of using a consistent base. In this way, even a very abstract-seeming calculation such as 2.3 × 4.72 can be seen as being essentially 23 copies of 472, the only difference being the units one is choosing to work with. This allows us to not have to worry about the meanings of the different place values while we perform the calculation and instead let the shifting do the bookkeeping for us.

Of course, it's even better to allow some common sense into the proceedings. Were I to calculate 2.3 × 4.72 myself, I would start by getting a reasonable estimate of what I should expect: I'm more or less multiplying two by five, so my answer should be in the neighborhood of ten, maybe a little more or a little less but definitely not in the hundreds or thousands, for instance.

Now I can ignore all the decimal point business and simply multiply 23 × 472 to get

$$8000 + 1400 + 40 + 1200 + 210 + 6 = 10856.$$

So, since I know it really needs to be around 10, it must in fact actually be 10.856. Alternatively, I could look at my original numbers and see that they are off from the numbers I multiplied by a total of three antishifts. Taking these into account, I again get 10.856. But I think the estimation idea is better.

While we're on the subject of decimal representation, place shifting, and estimation, it often happens (especially in scientific work) that one is presented with numbers—usually measurements of some kind—that are both astronomically large (or microscopically small) as well as *approximate*. In these cases it becomes convenient to take advantage of the place-shifting apparatus in order to save space. Instead of writing the number four billion six hundred million as 4 600 000 000, I could use the more succinct version 4.6 [9], meaning 4.6 with nine shifts. The approximate number of carbon atoms in a 12-gram sample is 6.02 [23] and the radius of a proton in meters is about 8.5 [−16] (meaning sixteen antishifts). This is a quick and convenient way to encode large-scale information in a compact form.

Calculate **110 × 101.1** *in binary. What numbers do these correspond to in the Hindu-Arabic system? How would the Banana People write this?*

DIVISION

Let's step away from the issues of representation and symbolic manipulation for a moment and think about what we are really doing when we are doing arithmetic. As I've said before (probably too many times), arithmetic is the art of arranging quantities for ease of comparison. This often takes the form of a collection of known quantities (however they may be represented or encoded) that we wish to combine in some way. For instance, we may have two amounts and wish to know their total. Fundamentally, we have two piles of rocks and we want to push them together to form one big pile. This is the essence of addition, independent of any representations or language. (And of course, this can be extended to as many piles as we wish.) The point being that addition is simply pushing piles together. Similarly, subtraction is the separation of a pile into two smaller piles, whose sum is then the original pile. In this way, we can see that adding and subtracting are inverse activities.

The modern view of arithmetic is that it consists of two categories of objects: *numbers* and *operations*. Numbers are the actors; operations are the actions. We certainly want to name our actors so as to be able to distinguish and compare them easily, but the content of the play is in what they *do*—the actions that the stage directions call upon them to perform. In the realm of arithmetic, numbers may be the nouns, but it is the operations that are the verbs. In some sense it is the operations that take center stage; the numbers serve only as a substrate for the operations to operate on. Perhaps I'm getting a little ahead of myself. All I mean to say is that we often find ourselves asking questions about numbers that concern their interplay, and it is the behavior of these various arithmetic activities (i.e., operations) that attracts our interest and curiosity. At any rate, this is the modern viewpoint—what matters is not the actors but the acting.

Addition, for instance, is an operation in the abstract (namely, pushing piles together) that does what it does independent of anyone's language or representation scheme. Of course, once you *have* adopted such a scheme, then it certainly behooves you to investigate what the operation of addition "looks like" in that system and to develop algorithms (and mnemonic devices such as needles) to make things quick and convenient. That's what we've been talking about this whole time. Be that as it may, addition is not columns of symbols and carrying; it's pushing piles together.

Clearly, addition is one of the simplest and most basic operations one can perform with numbers. What could be more fundamental than pushing two piles together? (Actually, as I've mentioned, there is one activity that is even more fundamental: comparison, the sine qua non of arithmetic. If we are not eventually going to compare quantities, then what is the point of reorganizing them? We may as well leave 173×248 alone.) Addition seems to be about the simplest thing you can do with two piles of rocks. And of course subtraction is really just unaddition, breaking the large total pile into two smaller ones.

Multiplication, on the other hand, is a bit more sophisticated. As an operation, we are taking two piles and using one of them to dictate how many copies to make of the other. Or, if you prefer, the two piles refer to the number of rows and columns of a rectangle of rocks, the total of which is their product. So multiplication is a bit indirect, a sort of meta-addition. And the fact that it happens to be symmetrical (e.g., $5 \times 6 = 6 \times 5$) should really be appreciated—not only as convenient but also as a surprising and beautiful feature of the operation itself (in keeping with the modern behavioralist viewpoint). In any case, multiplication is a perfectly nice operation—another verb in the grammar of arithmetic.

The thing about verbs is that whenever we have one, we always seem to get two. If I lock the door, then at some point I will need to unlock it. Tie a piece of string, and sooner or later someone will want to untie it. Actions that can be done almost always need to be undone. And this is especially true in mathematics, where symmetry is so highly prized and where the imaginary nature of the place allows us the freedom to reverse our actions so easily.

So we can view subtraction, for example, as an almost necessary linguistic construct. Once we can speak of addition, we then have the ability to ask questions: What do I add to this to get that? The desire to unadd follows immediately. What two-year-old, having learned to walk, does not eventually hit upon the perverse idea of walking backward?

Thus to every operation, activity, or transformation, we have the implicit possibility of an *inverse*, the reversal of the process. For every up there is a down; for every clockwise there is a whatchamacallit. To paraphrase the Book of Ecclesiastes, there is a time to cast away stones, and a time to gather stones together.

It so happens that making copies has a very natural inverse that arises quite frequently in arithmetic: *sharing*. Suppose you are lucky enough to be in possession of a rather large bag of chewy, mouthwatering jelly beans. Further suppose that you have somewhat foolishly promised to share them with your six so-called friends (or generously, if you want to put a brave face on it). The seven of you will receive equal shares to avoid the inevitable comparisons and accusations of favoritism.

The interesting aspect of such a sharing scenario is that we have no idea how many jelly beans each person should receive, only that we want the amounts to be equal. That is, whatever this mystery quantity is, seven copies of it should make the original total. This is the way that sharing is the inverse of multiplication.

This sharing business is technically known as *division* (from the Latin *divisus*, "separated"), and one is said to "divide" one number by another. In our case, we wish to divide the total number of jelly beans by seven. Let's suppose there happen to be two hundred forty-eight jelly beans in the bag. We want to know how to split it up—how to divide the jelly beans among the seven people. We know the number of copies, and we know their total. How do we figure out what each share should be? In short, what is 248 divided by 7?

One simple way to proceed would be to start handing out jelly beans, one to each person, going around in a circle in "eeny, meeny, miney, mo" fashion, until the bag (and you) are completely exhausted. The problem with this method is that it takes forever. Of course, if you only had a few dozen jelly beans, then it would actually be an ideal method— quick, easy, and very reliable. When the numbers get large, however, you'll be eeny-meenying yourself into an early grave. (Maybe *this* is why people don't like to share; communism will never work because the arithmetic is too annoying and time consuming?)

Luckily, there is a more efficient way. Instead of handing out jelly beans one at a time, we could hand them out two at a time, or three, or a whole pile. This will make the whole process go much more quickly (and may also restore our hopes for a communist utopia).

The danger here is that we will get too greedy. Not greedy for jelly beans but for efficiency. The more we hand out, the faster it will go, but we run the risk that we will give out too many too soon and run out before we get to everyone. That's bound to lead to trouble. In my experience, if there's one thing people don't like, it's giving back jelly beans.

This puts us in an amusing and interesting predicament. We want to distribute as large a handful to each person as we can, yet if we go too far, we may find ourselves "holding the bag," as they say. This means that a necessary part of any efficient sharing method is the ability to *estimate*—to get a rough idea of how many we can safely afford to hand out so that we don't run out too fast.

The other amusing (and possibly quite annoying) issue with sharing is that we often find ourselves with a few leftovers. After distributing the same number of jelly beans to all seven people, we may be left with two or three jelly beans in the bag. In other words, division doesn't always come out evenly.

What to do with the leftovers will of course depend on the circumstances. Let's say that in our case, to be fair, we will feed any leftovers to the dog. The general problem then becomes how to efficiently determine the division of the total, as well as the number of leftovers, if any. Incidentally, the number of leftovers is usually called the *remainder* (from Latin *remanere*, "to stay back").

In the case of jelly beans (or rocks), it probably makes the most sense to simply pour them out on the table and start making seven piles. We could start with a dozen or so in each pile, and then keep handing out five or so until the source pile gets smallish, then switch to maybe two each and then eeny-meeny it at the very end, until we see that we have too few left to give. These will constitute the remainder. Of course we could get lucky and have no leftovers. That's always very satisfying (unless you happen to be the dog).

A similar process would clearly work for a marked-value system, such as Egyptian counting coins. In practice, you may not have the actual objects to be distributed, or they may be too awkward to handle (e.g., goats). So you substitute counting coins for the actual objects and divide these instead. The point being that it is the *numbers* that we are interested in. You can always turn them back into goats at a later date.

Since you happen to be an ancient Egyptian goatherd, you may as well get some practice right now: a total of ९९९ ∩∩∩ ||| goats are to be divided among five goatherds. How many goats does each receive, and are there any leftovers?

Here we can imagine making five piles of coins, one for each goatherd. We could start by giving each pile a ९ coin and a ∩ coin, since we clearly have enough to go around:

$$९∩ \quad ९∩ \quad ९∩ \quad ९∩ \quad ९∩$$

This leaves us with ९९ ∩ ||| to distribute. Not having five of anything, it's time to do some arithmetic in the form of cashing in, rearranging the form of the number to make it easier to work with. Cashing in each ϙ coin in exchange for ten ∩ coins, we get a total of twenty-one ∩'s to spread among the five piles. Here is an opportunity to either be clever (e.g., by noticing that twenty is four fives and therefore also five fours) or prosaic (eeny-meeny until you're done). Either way, our piles now look like this:

ϙ 𐤀𐤀𐤀   ϙ 𐤀𐤀𐤀   ϙ 𐤀𐤀𐤀   ϙ 𐤀𐤀𐤀   ϙ 𐤀𐤀𐤀

and we are left with ∩ III. Finally, cashing in the ∩ coin for ten I coins, we can give each pile two and be left with three. So each goatherd gets ϙ 𐤀𐤀𐤀 II goats, and they can celebrate their successful division with a delicious three-goat stew. (This may also provide leftovers of a different sort.)

> *The pharaoh wishes to divide* ϙϙϙ 𐤀𐤀𐤀 \\\\\\\\\\\\\\\ *jewel-encrusted*
> *bracelets equally among his three daughters. How many*
> *does each princess get? Are there any leftover bracelets?*

The hallmark of our division procedure is its iterative nature. At each stage of the process, we have a certain number of items to be distributed among a given number of piles. We then estimate roughly how many we can afford to give away (keeping our estimates on the low side to be safe), and then we subtract these from our total and repeat the process. The larger our estimate (i.e., the amount we give to each pile) the faster everything will go but also the greater the danger will be of running out too soon. If, on the other hand, our estimates are too low, we will certainly be safe but probably also sorry that it is taking so long.

Let's see how this plays out in the Hindu-Arabic system. We'll use our jelly-bean example from before. What happens when 248 jelly beans are divided among 7 people? Rather than drawing pictures of piles, let's try to be as simple and efficient as possible. We'll need to keep track of a few things—namely, how many we have given to each person and how many are left to be handed out. So let's make a little table:

Amount Given to Each	Number Remaining
	248

Our table begins like this. And now we hit our first snag. We don't actually have in our hands any objects to arrange and distribute. This is one of the chief advantages of things like rocks and counting coins. Here we're going to somehow have to distribute things mentally and symbolically.

Not only that, but our estimates and our totals are going to have to be calculated symbolically as well. We will have to somehow—just by looking at the symbol 248—come up with an idea of how many we can afford to give each of the seven people, as well as knowing how many we will be giving away in total so we can deduct this from our original amount. How can we best go about doing this?

My advice (as always) is not to work too hard. Whatever amount we decide to give everyone, we're clearly going to have to multiply it by 7, so we should choose amounts that are easy to multiply. My suggestion would be to use only single-digit numbers like 8 or 400. Then we can let shifting do the work for us. We may possibly err on the side of giving away fewer than we could, and it may consequently take a bit longer, but at least we won't have to fill up mounds of scratch paper with a bunch of side computations.

Of course, even to use this lazy man's method requires that we have *some* sort of feeling for roughly how big various single-digit products are. But that's exactly the kind of thing, like language fluency, that comes from practice (and lots of it). For instance, I happen to know that $7 \times 3 = 21$, not because I burned that into my memory with drills and flashcards but because I care about numbers and am interested in their properties. Over the years, information like that eventually just sticks, especially if it is useful for my lifelong pursuit of entertainment.

Returning to our division project, I see that we have 248 jelly beans to distribute among 7 people, so I would like to find a simple quantity (i.e., a single-digit number) that when multiplied by 7 is more or less 248—actually, not more or less, just less. I want it to be as large as possible while still being safely below 248, and I don't want to work very hard, either.

Here is where my knowledge and experience with multiplication comes in. Clearly, I can't afford to give away a hundred to each person, because that would mean giving a total of seven hundred, which I ain't got. So, since I can't give away any hundreds, I'll try to give away as many tens as I can. Well, what if I give everyone ten? That would be a total of seventy, which I can certainly afford. Let's say we do that. Then we'll need to record the fact that we've given everyone ten, as well as figure out how many we have left. Here's a convenient way to do that with our table system:

Amount Given to Each	Number Remaining
	248
10	70
	178

After writing down 10 in the first column, I multiplied that by the seven people to get 70, and aligned it with the 248 so that I would be all set to subtract in the usual way. Now I see that each person has 10 and there are 178 jelly beans left in the bag. Does that all make sense? (Obviously, you can invent your own format for how you want to keep track of all this information; this way seems reasonably simple and straight-forward to me.)

Notice that we were not terribly efficient here with our estimate. We could have given away a lot more and made more of a dent in the total. Not that it's a big deal in any way. Truth be told, we could have given thirty jelly beans to each person and still been safe. Since $7 \times 3 = 21$, shifting tells me that $7 \times 30 = 210$, still well below 248. So if we had been a trifle more courageous, we could've gotten rid of a lot more jelly beans in one go.

There always seems to be a kind of "conservation of energy" principle with things like this. We can do a lot of simple arithmetic over and over, or we can work a little harder mentally but get it over with faster. (It reminds me a lot of pulleys: you can pull hard over a short distance and sweat like a pig, or you can invent a more elaborate compound pulley and have an easy—though lengthy—time of it.)

In any case, it's all water under the bridge. Let's not cry over spilled multiples. So we could have used 30 instead of 10, big deal. Let's just continue with our calculation. We have 178 jelly beans left. Any convenient multiples of seven lying around? Well, experience tells me that $7 \times 2 = 14$, so I know that $7 \times 20 = 140$, which is reasonably close and also safely under. So let's go with that:

Each	Remaining
	248
10	70
	178
20	140
	38

What our table tells us is that we have given away a total of $10 + 20 = 30$ to each person, and we have only 38 left. (This is where we would have been had we been clever enough to go with 30 in the first place.)

OK, so we have 38 jelly beans left. Clearly, giving even ten to each person is now too much. Of course, we could start handing out ones, but we can probably do a bit better. How about five? Since 7×5 happens to be 35, that's pretty much right on the money. So we end up with a total of 35 jelly beans for each person, with 3 leftovers. At the end, our table looks like this:

Each	Remaining
	248
10	70
	178
20	140
	38
5	35
35	3

Here I've added up the entries in the Each column to give me the total distributed to each person. (In this sort of calculation, it always pays to align the entries so that all the additions and subtractions are easy to perform and to read clearly.) Anyway, that's our division method in a nutshell.

What happens if we divide 5625 by 4?

Of course, people have found ways to streamline this process. We don't really need to label the columns, for instance; we could just agree to always put the Each numbers on the left and the Remaining on the right. And who's going to be looking anyway? If such a computation is going to be done at all, it will be done by you for your own purposes, so do whatever you like. Just be clear about what your symbols mean so you don't get confused.

More substantively, we can make our division procedure faster and easier by improving our estimates. In principle, we could try to maximize efficiency by always choosing the *largest* single-digit quantity we can afford so that there are fewer stages in the calculation. This is again a trade. To be faster and more efficient, we will have to be more knowledgeable and experienced.

In particular, we will need to have a fairly fluent command of the single-digit multiples. That's pretty much all that people mean when they say someone is "good with numbers." They mean that when such a person sees or hears "twenty-seven" they immediately think to themselves "three nines." Maybe it's a case of familiarity breeds contempt, but I actually don't think that arithmetic skill is anything to be all that proud of or envious about. If you care to develop that kind of skill, then you will. Walking and talking are infinitely more difficult, and most toddlers eventually seem to get it down. So get as good at it as you like, but please don't stress over it.

Anyway, let's suppose that we more or less know our single-digit multiples (i.e., the times tables). We need to know this information in reverse as well; not just that three sixes is eighteen, but that eighteen is three sixes. That's pretty much what *is* means—that two things can be used interchangeably. It's a bit like learning a foreign language; you need to be able to go from English to German and back again. And here we really are talking about a dictionary—we're translating from groups of six to groups of ten and back again. That's what the times tables really are: a dictionary from decimal to other grouping sizes. Anyway, let's suppose we know all that. Then we can speed up our method a bit.

The idea will be to give away all the thousands we can (making our estimates maximal), then all the hundreds, tens, and finally, ones. So there will be one stage of the process for each decimal place. Let's put it into practice with an example: 6842 divided by 5. Here is what it would look like in table form:

$$
\begin{array}{rr}
 & 6842 \\
1000 & \underline{5000} \\
 & 1842 \\
300 & \underline{1500} \\
 & 342 \\
60 & \underline{300} \\
 & 42 \\
\underline{8} & \underline{40} \\
1368 & 2 \\
\end{array}
$$

At each stage we used our intimate knowledge of multiples of five to always make the largest (and therefore most efficient) estimate. So if you like that sort of thing, it's there for you.

In fact, this process can be streamlined even further if you are so inclined. Since we are always making single-digit estimates, they are particularly easy to add up. Notice how we obtained the grand total 1368: it came from adding 1000 + 300 + 60 + 8. In other words, the single-digit numbers don't interfere with each other; there is no exchanging or carrying or anything like that. The digit sequence 1368 is just sitting there in the estimates themselves. Some people like to exploit this feature to make the algorithm even faster and to save space. One popular format is to put the single-digit estimates *above* the original quantity, aligned correctly with the corresponding place values:

$$
\begin{array}{r}
\underline{1368} \\
6842 \\
\underline{5000} \\
1842 \\
\underline{1500} \\
342 \\
\underline{300} \\
42 \\
\underline{40} \\
2
\end{array}
$$

The point here is that no adding needs to be done at the end; the digits just fall into place as you go. This can be a convenient time and space saver when the numbers get big.

Try the new method with 53219 divided by 8.

Of course, things can get complicated. For instance, not only might the number of jelly beans get quite large but also the number of people. You may find yourself needing to divide 17503 by 13, say. In this case both the estimation as well as the multiplication are going to get messy. Let's give it a try.

Our calculation begins simply enough, seeing that we can easily give away 1000 each. Let's make the usual table:

```
                          17503
        1000              13000
                           4503
```

Now we have to ask, how many hundreds can we afford? Knowing the multiples of thirteen would certainly help, but we can't know all multiples of all numbers—there are infinitely many of them! So sometimes we have to buckle down and do some side calculations. I know that three copies of 13 is 39. Can I afford four? I guess I can use the fact that a deck of cards has four suits of 13 cards each and is 52 cards total. (Don't laugh—all experience is grist for the arithmetic mill.) This means that 300 each will cost 3900 and 400 each would be 5200. So I can only afford to give three hundreds:

```
                          17503
        1000              13000
                           4503
         300               3900
                            603
```

So you see this division business can be a lot of work: estimating, multiplying, subtracting, and aligning everything—no wonder I avoid it whenever possible!

At this stage we need to estimate how many tens we can afford to give away. I wouldn't blame you (or myself) if we got lazy at this point and simply gave away 30 each, since we already know it's 390. This would spare us some multiplication, but we would then need to give away more tens later. On the other hand, we can still get away with doing no labor by employing the deck of cards again: 13 × 40 must be 520, which is larger but still affordable. So let's go with that:

```
                          17503
          1000            13000
                           4503
           300            3900
                           603
            40             520
                            83
```

Being complete slobs, let's just use 13 × 4 again for the ones, and then deal with the aftermath (no pun intended):

```
                          17503
          1000            13000
                           4503
           300            3900
                           603
            40             520
                            83
             4             52
                            31
             2             26
          1346             5
```

OK, so we could have afforded six ones at the end. In any event, we are done: 17503 divided by 13 is 1346, with a remainder of 5. The (slightly) more efficient version would appear thus (I've even gotten cute here and suppressed unnecessary zeros to save writing):

```
              1346
            17503
            13
             4503
             39
              603
              52
               83
               78
                5
```

Truth be told, I'm not actually all that big a fan of the more efficient version. I don't really want to work that hard to ensure that I'm always giving away the greatest possible amount each time. I'd rather just give out whatever I want to whenever I want to. In fact, sometimes I will even give away *too much* (that is, more than I can afford) and then take a little back later just to make my life easier with the estimation. So my advice is to do what you like, efficiency be damned.

The truth is, pencil-and-paper division is not a terribly common occurrence, and as we shall see, most of these sorts of lengthy algorithmic procedures are pretty much obsolete at this point anyway. But here is a problem to try out just for fun.

> *Four thousand six hundred forty-eight chocolate bars*
> *are to be divided into seventeen equal shipments.*
> *You have been promised the leftovers, if any.*
> *How many chocolate bars do you get?*

While we're on the subject of division and sharing, I want to mention an amusing feature. When we say that we are dividing a bag of jelly beans among seven people, we are asking what number, when multiplied by seven, makes the given total (let's ignore leftovers for the moment). In other words, seven times what makes a bagful? For definiteness, let's say that each person gets eight, so there are fifty-six jelly beans in all. The fact that 7×8 equals 56 is capable of multiple interpretations (another unintended pun!).

On the one hand, we can think of 7×8 as seven copies of eight (so that each of the seven people receives eight jelly beans). Alternatively, using the symmetry of multiplication, we can also view it as being eight groups of seven. So seven is the number of people to whom we could give eight jelly beans each. In other words, instead of asking how many can go in each pile, we can ask how many piles can receive a given amount. The symmetry of multiplication gives us two slightly different ways to think about division: we are chopping a quantity up into a given number of piles whose size we seek, or we are chopping it into piles of a fixed size and we want to know how many piles there will be.

Of course, these are really the same in the abstract; the issue is really one of interpretation. Neither seven nor eight (nor their product, fifty-six) cares about you and your arithmetic problems. They are too busy doing what they do (namely, multiplying) to bother with your petty interpretation issues. The modern view would be to ignore piles and shares and people and just focus on behavior: fifty-six divided by eight is seven because seven times eight is fifty-six, period. Who cares which number is thought of as people and which one as piles?

The other thing I wanted to point out is that our division procedure gets along very well with place shifting. Just as multiplying by sixty is essentially the same as multiplying by six, so it is with division. In fact, since multiplication by ten can be thought of as simply a shift (one place to the left), division by ten can likewise be viewed as an antishift (i.e., a shift one place to the right).

This means that all of our division procedures survive the passage to decimals, that is, representations involving subdivided units and decimal points. For example, if a loan payment of $324.87 is to be divided into five equal installments, we could proceed in the following way. First of all, forget the decimal point. We have 32487 of something (in this case, pennies). Let's divide by five as usual:

	32487
6000	30000
	2487
400	2000
	487
90	450
	37
7	35
6497	2

Because the place shifting can be done at any time (i.e., you don't have to commit to what your units are), we save it until the very end. So here I could say that since I initially did a double shift to turn 324.87 into 32487, now I can undo that to see that each installment should be $64.97, with $.02 remaining—in other words, there are two leftover pennies. (You'll have to work that out with your lender.)

The point is that we have a lot of flexibility with these place-value algorithms as long as we can keep a clear head.

> *What happens if we are to pay the* $324.87
> *in fifty equal installments?*

It often happens, especially in more technical arenas such as science and engineering, that we require a higher degree of accuracy in our computations. In such cases remainders become problematic. For example, if I have an 8 liter sample of liquid weighing 7.25 kilograms (that is, 7250 grams), the density of the liquid (measured in grams per liter) would be calculated by division: we want 7250 divided by 8. Running the usual procedure gives:

$$
\begin{array}{rr}
 & 7250 \\
900 & 7200 \\
\hline
 & 50 \\
6 & 48 \\
\hline
906 & 2
\end{array}
$$

But here the leftovers cause trouble. Yes, we could say that the density is roughly 906 grams per liter, but that's what it would be if the sample weighed 2 grams less. I suppose we could pour off a little and reweigh it, but then the 8 liters would be wrong. The correct thing to do with something like this is to subdivide our units more finely. Instead of viewing 7.25 kilograms as 7250 grams, let's think of it as 7250000. (This would mean we are measuring in *milligrams*, but it doesn't really matter. The point is that we are adding on a bunch of shifts to increase the accuracy.) Now we get:

	7250000
900000	7200000
	50000
6000	48000
	2000
200	1600
	400
50	400
906250	0

and there are no unpleasant remainders. Shifting back—to return to our original units—we obtain a density of 906.25 grams per liter.

Had there still been leftovers, we could either shift again (to get even more accuracy), or we could simply ignore the remainder if it represents an amount so small that the error is negligible. We may not care about being off by .00003 grams per liter, say.

In every practical, real-world scenario (including even particle physics and cosmology), there is always going to be some threshold of accuracy beyond which things become pointless. For one thing, your original data is always approximate anyway. So scientists and engineers, carpenters and seamstresses, bakers and farmers all use arithmetic frequently, but the most important skill is knowing what degree of accuracy is appropriate—in particular, when remainders can safely be discarded. If there is such a thing as an accepted convention, it would be this: *don't bother being more accurate than your data.* If the situation calls for you to divide 125.26 by 3.8, it would be absurd to work hard to produce a figure like 32.9631579, when the original numbers are only given to one or two places after the decimal point. I would probably just go with 33, seeing as it's so close.

Generally, when someone writes a number like 3.8, what

they mean is that all they can guarantee is that it's *around* 3.8, somewhere between 3.75 and 3.85, say. If more were known (or cared about), then maybe we would have a more accurate estimate like 3.83, which would again only tell us the number up to a certain tolerance—the final digit always being a bit iffy. The decision of how accurate one's measurements should be, and thus how many so-called significant figures to provide, is a very basic one. It really just depends on what you are doing and why.

You are in the laboratory measuring the amount of liquid contained in three test tubes. The volumes (in milliliters) are found to be: 30.25, 27.12, and 32.62. After combining them, you find the total volume to be just a hair over 90 milliliters. Should you be surprised?

MACHINES

Perhaps the most surprising and powerful aspect of place-value arithmetic is how it reduces any calculation to a set of purely abstract symbolic manipulations. In principle, I suppose, one could even be trained to perform such symbol-jiggling procedures *without any comprehension whatever* of the underlying meaning. We could even (if we can possibly imagine being so cruel) force young children to memorize tables of symbols and meaningless step-by-step procedures, and then reward or punish them for their skill (or lack thereof) in this dreary and soulless activity. This would help protect our future office workers from accidentally gaining a personal relationship to arithmetic as a craft or enjoying the perspective that outlook would provide. We could turn the entire enterprise into a rote mechanical process and then reward those who show the most willingness to be made into reliable and obedient tools. I wonder if you can imagine such a nightmarish, dystopian world? Let's try not to think about it.

Whenever humans start to realize that they are performing mindless repetitive tasks, someone eventually has the bright idea to build a machine to do it instead. Arithmetic is no exception. Why should I waste my time moving symbols around in a purely rote mechanical way when I can get my servant to do it for me? And why should I put up with a cranky and unreliable human servant (who might even expect to be *paid*) when I can employ an inanimate object to do my work for me? Even the human machines that we produce in our obedience schools are no match for a mechanical slave.

From the plow to the potter's wheel, from the loom to the sewing machine, humans have always found ways to mechanize useful activities, transferring the hard boring labor over to the machines, freeing us up to think and relax and make art. (It is also a great way to eliminate and impoverish entire swaths of the industrial labor force and create widespread inequity, social unrest, and eventual class warfare.)

The simplest mechanical arithmetic device would be a *counter*, such as an odometer or a turnstile, where each turn of the wheel causes the counter to increment. There are simple handheld devices that are used to tally: when a button is pushed the number increases by one. How do such machines work?

The trick to getting a machine to do something for us is to find a way to remove comprehension and understanding from the process; to reduce it to a purely mechanical activity, where not only is it unnecessary to understand what is going on, but it is *essential* that there be no understanding needed, else how are we going to get a bunch of mindless gears and springs to do it?

Which is not to say that machines are in principle unable to think, only that we don't yet know exactly what thinking is, or how to make it happen (though we seem to have gotten pretty good at finding ways to make it *not* happen). Of course, there are biological machines that are apparently able to think—us, for example—but we didn't design them and we don't really know how they work.

So the first step in the construction of any machine is to break down the procedure you wish it to perform into a sequence of mindless, rote mechanical activities. When we are weaving a tapestry, we are well aware of how we are choosing a green thread to make part of a leaf design, but an automated loom can have no such ideas; it must simply follow instructions uncomprehendingly, selecting a green thread not because it intends to but because it is *forced to* by its own design. Similarly, I don't choose to have blood course through my veins; my heart beats whether I like it or not. (For the most part, I'm glad that it does.)

So we have to somehow get a mechanical counter to "know" how to increment (that is, add one to the count) without really knowing. In particular, although the device may *display* Hindu-Arabic numerals for our benefit, it will not (and cannot) understand their meaning. That means we need to come up with a physical interpretation of "adding one" that is inherently meaningless.

One simple idea would be to make a scroll—a long strip of paper with consecutive numbers written on it—so that each turn of a crank, say, would advance the scroll to the next higher number.

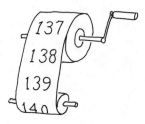

This is certainly mindless enough. The deluxe version would probably include some sort of a housing or box to put it in to make it easy to hold and to operate. We would also want some kind of display window so we could see which number we've gotten to. Finally, we'd probably like to engineer some way to keep it from being turned accidentally, or slipping backward, so that it isn't too fragile and finicky. Maybe a satisfying click each time we advance the scroll would be nice as well. The typical solution is to include a ratchet mechanism. This is simply a gear, together with a spring-loaded pin to keep the crank from turning too freely.

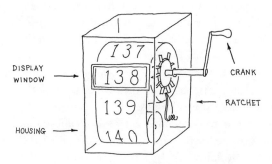

Though almost laughably simple, this tallying device has some very convenient features. It is certainly easy to use: just turn the crank! The current count appears in the display window. With very slight modification to the ratchet, we could even allow it to turn backward, if that were desired, to correct for accidental overcounting, say.

Sooner or later (maybe now, even), you may start to notice some inconveniences inherent in this design. Quite apart from the annoyance of having to prepare a long scroll of consecutive numbers in the first place (and how long would depend on what you intended to use it for), there is the problem of having to reset it, or "zero it out," each time you want to use it. If yesterday's tally made it to three hundred and something, then in order to reset the device for today's count, you would have to spend some time rewinding the scroll back to the start (which would presumably display the number zero). That's pretty annoying.

Can we do better? I suppose we could have a ready supply of fresh scrolls and switch them out each time, but that hardly seems best. I'd prefer not to open the thing up and tinker around if I don't have to. I just want a reliable, easy-to-use handheld device—and I don't really want to make a long scroll of numbers either, if I can avoid it.

The thing about technology is that it evolves. Just as plants and animals respond to the changing vicissitudes of the environment, flourishing or suffering depending on the aptness of their design, becoming extinct or mutating into better-adapted forms, so it goes with the machines. The environment is human usage, and innovation drives the evolutionary process. Once a new idea is discovered that corrects a problem or eliminates an annoyance, it's curtains for the old design. Museums are filled with dinosaurs of both the animal and mechanical kind. The big difference is the rate of evolutionary change. Technology evolves much faster than natural organisms because our innovations are intentional, as opposed to a random, directionless process of genetic mutation. When we wish to improve a feature of an existing mechanical design, we do it on purpose, with a goal in mind. We don't always succeed, and we are notoriously bad at foreseeing unintended side effects and negative consequences, but it is far from random. This means that technological change is both rapid and unpredictable, but what are you gonna do? Humans are insatiably curious, remarkably clever, and unbelievably stupid.

Hopefully, improving our tally machine won't cause too much unintentional harm. The big historical innovation is the idea of a *digit wheel*. Instead of a cumbersome written scroll of numbers, we simply use a separate wheel and axle for each digit. This not only frees us from having to make a long list of numbers but also allows us to affect and control each digit separately, making it much faster and easier to reset.

Each digit wheel will require its own ratchet mechanism, and there will be further complications as well, as we will shortly see. This is quite typical of technological innovation and evolution. We certainly saw it with the history and development of number languages. Almost always it comes down to a trade: easier to use, harder to understand; faster and more convenient, less intuitive; has more features, breaks more often.

The problem with evolution—whether mechanical, biological, or cultural—is in knowing when to stop. Were eighty-foot lizards really such a good idea? How about high heels and TV remotes? Which innovations are really useful and which are merely fads that later generations will find ludicrous?

I think the digit wheel is a good invention. There are many variations on the basic idea, but the one I like best looks like this:

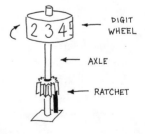

In this design, each time the wheel is turned on its axle, the ratchet makes a click and the front-facing digit is advanced.

For a multidigit counter, we would simply arrange several of these digit wheels in sequence.

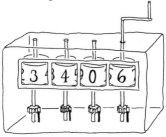

Here is a design for a four-digit counter. Each digit gets its own display window, and the crank (or knob, if you prefer) can be attached directly to the axle of the ones digit wheel.

You may already have noticed our first snag. When the counter reaches the number nine, and we then give the crank another turn, instead of going on to ten, the counter will reset to zero. The trouble is that the ones digit just spins around; we need to give it some way to inform the tens digit that it has just "rolled over" to zero, and it is time for the tens place to increment. So the snag is that there is no actual *snag*—we need to get the ones digit wheel to catch onto the tens digit wheel at just the right time.

The simplest way to do this is to attach a "carry pin" to the ones axle and a corresponding "carry gear" on the tens axle, like so:

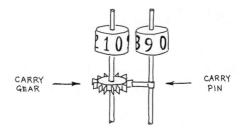

The carry pin simply rotates along with the ones digit wheel, doing nothing at all for most of its journey, until the wheel gets to the all-important 9 position. At this point, the carry pin engages with the carry gear on the tens axle so that when the ones digit advances to the 0 position, it automatically forces the tens wheel to rotate as well. Of course, the width of the pin and the spacing of the gear teeth will have to be designed with great care so that the tens digit wheel will advance just the right amount, but that's the kind of nitpicky detail that engineers drool over.

There is a curious side effect of this plan, actually, which is that the rotation of the ones digit wheel from 9 to 0 will cause the tens wheel to rotate *the other way*. It is the nature of interconnected gears to rotate oppositely. If the ones wheel (and thus the carry pin) is rotated clockwise, the tens wheel will necessarily rotate counterclockwise. That means we have to reverse the order of the digits on the tens wheel, as I have done in the drawing. For some reason I find this rather amusing. Of course if you don't like it, you can always try to engineer your way around it (e.g., by inserting more gears and so forth), but let's keep things simple.

While we're at it, another improvement we can make is not only to replace the crank with a less cumbersome knob or dial but also to install one of these on each of the digit axles to make them independently adjustable. Our new, improved counting device might look something like this:

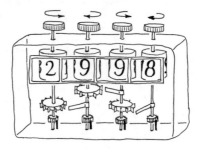

Each axle is provided with a carry pin as well as a carry gear (though the axles at each end do not require both). Since these can be placed arbitrarily high or low along the axles, they can be staggered (as shown) so as not to interfere with each other. The wheels will then rotate in alternate fashion, clockwise and counterclockwise.

At last we have a counting machine worthy of the name. Also, since each axle has its own control, we can easily add twenty to the count simply by turning the tens dial twice (in the proper direction, of course). Not only that, but it is also easy to subtract by turning the dials the opposite way.

When I was a kid and my mom took me to the market, she used exactly this kind of device to keep track of how much she was spending. Often she would let me hold it and would tell me the price of the next item, say $2.39. Then I got to turn the dials (or were they buttons?) to add it on to the total: two turns of the hundreds wheel, three for the tens, and nine for the ones place. It simply delighted me when a digit rolled over and advanced the next place automatically. I loved it when I was six, and I love it now. Can anyone resist watching an odometer roll over to 100,000 miles? What exactly is the universal appeal of the carry pin?

I think it is the essence of life itself. Inanimate objects are somehow intrinsically boring compared to things that move. When something moves by itself, we like to watch. Not only that, but we tend to endow that thing with the status of a fellow being. We say, "Look what it's doing" and "What does it want?" and things like that. We ascribe to it a certain level of consciousness and intentionality. If a thing is complex enough, we may even grant it a personality, as in "My car has been really cranky lately" or "The toaster doesn't seem to like me today." This is also borne out by the etymology of words like *animate* and *animal* (from the Latin *anima*, "soul"). When things move by themselves, requiring no impetus from us, we tend to view them as living creatures.

Perhaps the most crucial requirement is that the thing *reacts*—it does different things depending on external circumstances. Just because water flows along a riverbed doesn't make it alive. We want to see it adjust its behavior and "make decisions."

In an admittedly laughable and simplistic way, this is exactly what a carry pin does. It makes our counting machine appear to understand something about place-value arithmetic. The tens wheel is just sitting there, minding its own business while we turn the ones dial, until we pass 9 and get to 0, at which point the tens wheel seems to react to that news and moves, in a somewhat creepy way, *all by itself*.

Perhaps you roll your eyes and scoff at my fanciful notion of endowing a counting device with a consciousness (of whatever limited form). But don't people often do something like this with computers? We may understand intellectually that such machines (which are obviously quite a bit more complex than our mechanical counter) are nothing but a massive pile of interconnected transistors, yet we still find ourselves saying, "What is it up to?" and "I wonder why it decided to erase my file?" and so forth. Fundamentally, a transistor is a switch: it transmits electrical current or not depending on a certain input voltage. That is, it reacts differently to different circumstances. So it's really just a fancy carry pin of sorts. Hook enough of them together, and you get a "thinking machine."

For that matter, what is a neuron but a (ridiculously sophisticated) biological carry pin? Neurons also send electrical impulses depending on their inputs, just like transistors—the major difference being that we didn't invent them. Consequently, we don't yet entirely understand how they work. Nevertheless, a neuron is a biological switch all the same, and if you connect enough of them together (say one hundred billion or so), you get a person's brain. And we don't seem to have a problem endowing *those* things with intentionality and consciousness.

So I wonder if it's really so silly after all to feel a certain animal response when we watch a counter roll over from 0999 to 1000. Who's to say that's not what we are all doing, more or less, when we sit and talk with a friend over a cup of coffee? A few million neuronal carry pins in my brain have just triggered my (apparent) desire to take a sip of coffee and to quickly glance at my watch and say, "Yes, I completely understand your feelings." I don't know about you, but I'm not too proud to accept the idea that I'm nothing more than a complicated mass of carry pins. Which is pretty funny, given that I'm sitting here talking to you about them.

In fact, the whole world is pretty much a case of runaway automation. Everything's just moving around, reacting to the carry pins of chemistry and physics. So laugh all you want. I'm sticking with my six-year-old feeling of delight in things that move by themselves. You don't want to think of an odometer as alive and possessed of a soul, then don't. But Odometer and I have a beautiful relationship, and when she rolls over, I go to pieces.

In any event, we have a simple and reliable mechanical device that automatically keeps count. Using the individual place value dials, we can easily add any quantity to the current total and the machine (via its carry pins) will perform the necessary "exchanges" despite its complete lack of comprehension.

As you might imagine, people have elaborated on these ideas considerably over the centuries, from Leonardo's fifteenth-century wooden gears and carry pins to modern high-precision machine parts. Design improvements not only allowed for increased reliability and ease of operation but also expanded the range of operations to include addition, subtraction, place shifting, multiplication, and division. Naturally, these require considerably more elaborate mechanical realizations, but in principle it should be in no way surprising. If a task can be performed in a completely rote, step-by-step manner, it's only a matter of time before some clever engineer finds a way to get some gears and levers to do it as well.

The epitome of handheld mechanical calculating devices was attained in the late 1930s with the development of the Curta calculator.

Designed by an Austrian engineer (and concentration camp inmate) during World War II, the Curta was (and still is) the most reliable, convenient, and powerful mechanical calculating machine of its size. Accurate to as many as fifteen decimal places, the device was capable of performing any multidigit addition, subtraction, multiplication, or division in a matter of seconds.

Of course, it wasn't long before fully automatic electric versions made their appearance. I remember playing with my dad's electric tabletop calculator when I was little. It was large, green, heavy, and most of all, *loud*.

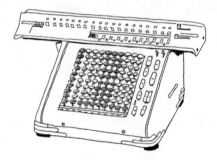

To multiply two numbers, you would push digit keys for each decimal place, press the multiply button, and—*ka-chunk-a-chunk-a-chunk*—it would start place shifting and totaling. A few seconds later the product would appear in the display window. You didn't even have to turn a crank!

The electric mechanical desktop calculator (along with

its cousin, the automatic cash register) pretty much meant the end of old-fashioned pencil-and-paper arithmetic. Sure, people still do short back-of-the-envelope computations now and then, but for most large-scale commercial or scientific calculations, we let the machines do the work, as well we should. After all, they're *better* at it than we are. (These days, the only people who actually bother to carry out the pencil-and-paper algorithms are schoolchildren, and only because they are forced to.)

The real inconvenience of these machines was their size and weight, making them far less portable than a soroban or a pencil. This final technical hurdle was overcome in the early 1950s with the invention of the transistor, a solid-state electronic device that essentially acts as a voltage-controlled switch, replacing the older and more cumbersome vacuum tubes of my childhood. (I can still remember how exciting it was when my dad would let me use the tube tester at the local drugstore to check for faulty radio and TV tubes.) Almost immediately, smaller, more lightweight and portable transistor radios appeared, allowing people for the first time in human history to have music wherever they go.

The transistor, like the vacuum tube before it, is a *triode*, a device that allows current to flow (or not) in a definite and controllable way.

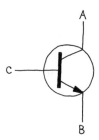

Here, in this schematic diagram, current will flow between terminals A and B, but only if the third terminal C is set to the correct voltage level. Otherwise, the current will not flow. This allows us to design and build electronic circuits that act "intelligently": like carry pins, they behave differently depending on the circumstances.

In particular, we can create logic and memory circuits that can hold and manipulate information depending on, for instance, which buttons are pressed. In this way, the mechanical calculator can be completely redesigned electronically, with no noise and *no moving parts at all* (except electrons, which are notoriously quiet). The role of the carry pin is taken over by the logic circuitry, and the digit wheels and display window are replaced by the (now ubiquitous) seven-segment light-emitting diode (LED) display pictured below.

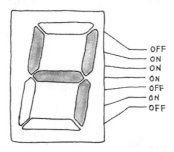

Each terminal supplies current to a separate LED. The various digit symbols then correspond to their own pattern of input voltages (which, for simplicity, I've labeled **ON** and **OFF**). Thus, the digit 4 is displayed as shown with the pattern

OFF ON ON ON OFF ON OFF

There is something rather amusing about this situation, which provides yet another example of numerical representation. First of all, because of the simple two-state nature of digital circuitry—voltages can be either **ON** or **OFF**—it is typical to encode information in *binary* (i.e., base two). The number thirteen, for instance, would be held in a calculator's memory as the sequence **ON ON OFF ON**, corresponding to the binary representation **1101**. Then all of the logic circuitry for addition and so forth must be designed accordingly. That is, we add numbers in base two. Finally, in order to please the conventional hidebound human operators (who want to see the usual base-ten digit symbols), the circuitry must then take the additional step of converting this information into a completely different ON/OFF sequence in order to drive the display LEDs.

Of course, it is no small matter to design and build all of these circuits and to get them to behave properly, but that's what makes electronics such challenging fun. And talk about seeming alive—even the "take a number" thing at the deli increments automatically and without any gears at all.

By the time I was in high school in the late 1970s, the electronic pocket calculator was everywhere—fast, silent, cheap, and dependable. Easy to use, lightweight, and portable, these devices instantly made all previous calculating technology obsolete. Now there's one in every kitchen junk drawer. Continued miniaturization and mass production have made them not only faster and more convenient than pencil and paper but also cheaper and lighter as well.

These days, even the most powerful handheld electronic calculators are essentially extinct (surviving only by virtue of the bribes paid by the manufacturers to the educational testing companies), having been replaced by software versions on multipurpose appliance computers such as laptops and smartphones. Now you can perform a calculation instantly on your wristwatch that would have taken Archimedes months to scratch in the sand.

This is the way of all technology, of course. What begins as a sophisticated human craft requiring years of apprenticeship and experience is gradually refined and reduced to a series of repetitive steps to be performed mindlessly and eventually relegated to a machine. This has the positive effect of freeing the craftsman to put his energies into the more creative aspects of the work. A painter has more time to work on the composition and lighting effects if she doesn't have to manufacture her own paints and brushes.

The downside of all of this mechanization is that it allows for the operation of the devices without the understanding and sophistication of taste that comes with the older artisanal approach. Growing up in a print shop not only exposes you to a variety of typefaces and graphic design options, but the hand-setting of each line of type, the demands of registration and alignment required by a hand press, the smell and texture of the ink under your fingernails creates a culture and an appreciation for the aesthetic fine points that is clearly and woefully lacking in the modern point-and-click world of automated typesetting.

So I guess the question comes down to one of goals and practicality. If you want to produce printed works of impeccable taste and quality, I suggest you apprentice yourself to an Old World printer and get out the composing sticks and chase furniture. If all you want is to print out the office party flyer, then I guess you might as well just grab the mouse and click on "flyer template," or whatever.

Similarly, if all you want out of arithmetic is to total today's receipts or to do your taxes, then by all means get out the calculator. I certainly would. There really is no practical reason why anyone needs to understand arithmetic on a deeper level unless he or she wants to. For myself, I could never be satisfied with using a system or a machine without understanding how and why it works. My curiosity is just too strong. I step on a pedal and the car moves by itself. How on Earth does that work? (I think I'd rather take a car apart than use it to carry groceries.) In the case of arithmetic, not only do I want understanding for its own sake—"because it's there"—but also because as a mathematician I'm interested in numbers and how they behave, both in the abstract as well as in symbolically encoded form. That's why I especially want to be independent from any particular cultural choices and to be as flexible in my thinking as possible. I want the perspective and the intellectual and creative power that comes from an abstract understanding of arithmetic, as well as the convenience of fast, reliable, inexpensive, and lightweight electronic calculators. Give me a machine for the dull mechanical grunt work, and leave my mind (that *other* machine) free for the more interesting and imaginative things in life, such as mathematics.

FRACTIONS

So the practical problem of arithmetic has been completely and utterly solved. Virtually every human being now has access to a cheap, portable, and absolutely reliable electronic calculating device, and a great many carry such things on their persons wherever they go. The long historical saga of representations and abacus systems has come to an end. So you can take your tabula and your calculus stones, your counting coins and soroban and throw them all away. Leave the pencil and paper in the drawer. If you need to do some Hindu-Arabic decimal place-value arithmetic, simply take out your phone and push some buttons.

There is no cause for weeping and moaning; it is simply the way of all things. Where is your family spinning wheel? Who wove that shirt you are wearing? I'll tell you who: a massive automated programmable industrial textile loom, that's who. But it's no great loss. You can still knit and spin the Old World way, if you want to. It will take a lot longer (and will probably cost a lot more), but you can go buy yourself a skein of wool and some knitting needles and have a grand old time. In fact, I recommend it.

Similarly, you can still do arithmetic by hand if you want to. Piles of Rocks is still there as an option, and you can always make your own tabula (if you haven't already). Many older Japanese still prefer using the soroban, and are remarkably quick and skillful at it. Perhaps this may even rekindle your relationship to pencil-and-paper arithmetic—seeing it anew as a quaint, outmoded thirteenth-century folk art. Whatever. The point is that the story of arithmetic, at least from a mundane practical point of view, is over. Now we can turn our attention to the more conceptual, theoretical aspects of this beautiful subject. Our goal is not utility or practicality but intellectual pleasure and understanding.

I want to return to an issue that came up when we were talking about division—namely, what to do with those pesky remainders. Let's say we have seventeen jelly beans we wish to divide among four people. Certainly we can give each person four jelly beans, and then we are stuck with one leftover. Before, we simply fed it to the dog. Now I want to look at a different option.

Suppose we get out a sharp knife and actually *chop up* our remaining jelly bean into pieces. If we are careful, we can cut the jelly bean into four pieces of the same size, and then give each person one of the pieces. The upshot is that we have *completely* divided the jelly beans among the four people, with no leftovers at all. Each person receives four whole jelly beans and one (possibly sticky) partial jelly bean, known as a *fraction* (from the Latin *fractus*, "broken"). We would say that the leftover jelly bean was quartered (i.e., chopped into four equal pieces) and that each person gets four and a quarter jelly beans. So, depending on the situation, we have the option of dividing completely (using fractional pieces if necessary) or sticking with our old system of remainders (and possibly dogs).

Of course, not all objects are equally amenable to this chopping procedure. If we were dividing up a collection of violins, for example, there's not going to be any chopping into pieces. Some things allow for that and others don't. It has to do with whether we consider them to be irreducible or not—what I like to call "lumpy" versus "smooth." Violins are quite lumpy; if you cut one in half, it is no longer a violin. The same goes for dogs. Milk, on the other hand, is happy to be subdivided into as many parts as you like. If we have a gallon of milk to share among four people, then each person gets a quarter gallon (or quart, for short). If we have more people, then we might have to resort to pints, cups, tablespoons, and eventually drops, but the milk does not put up a fuss or cease being milk. Milk is smooth, as are all other liquids.

The point being that when you are dividing things, it is important to understand whether they are lumpy or smooth. Do they come in individual, indivisible units, or are they capable of arbitrary, unlimited subdivision?

Of course, in real life, this decision can get rather tricky. There is a long history of philosophical and scientific debate concerning the nature of reality—is matter smooth, or does it come in lumps? (The word *atom*, in fact, comes from the Greek for "indivisible.") And what about time and space? The truth is, we still don't know the extent to which real objects can be chopped. A chemist might say that an atom of gold is still gold, but once it is split up into its constituent subatomic particles, it ceases to be gold. In that sense, gold is lumpy. On the other hand, an *ounce* of solid gold (which consists of billions upon billions of individual atoms) is, for all practical purposes, perfectly smooth. So it really depends on the context in which you are operating. Things are lumpy or smooth depending on how you intend to treat them. Do you allow them to be chopped into fractional pieces or not?

Make a list of things that you think of as smooth or lumpy.
What are some things that can be treated both ways?

The point is, if we are comfortable subdividing our individual objects, then we can always divide completely and avoid any unpleasant remainders. Let's imagine we are in such a situation. In fact, let's say that we are ancient Egyptian goatherds dividing up a quantity of goat milk. Let's suppose we have 𓏏𓏏𓏏 III jars of milk to share among the four of us.

Well, we can each take ∩ jars and still have ∩ III jars remaining. Then we can each take III more, and have one leftover jar to split between us. That means we should each receive an additional one-quarter (what the Egyptians called "the fourth part") of a jar. This was written 𓏺, the lentil-shaped symbol indicating "part." Similarly, one-half could be represented 𓏻 (although for the more commonly used fractions such as one-half and two-thirds, there were special shorthand versions as well).

If instead we had twenty-seven jars of milk (that is, ∩∩ 𓏼𓏼), then after giving out six jars each to the four of us, we would be stuck with three remaining jars. How do we divide three things among four people?

The Egyptian method was to simply continue doling out milk, but instead of asking how many tens or ones we can afford, we switch to subdivided units and ask how many halves we can afford, then how many thirds, quarters, and so on.

Here we have three jars to share among four people. We clearly cannot afford to give away a whole jar to each. How about giving away half jars? If we give each of the four goatherds half a jar, that would make two full jars with one jar still remaining. So then we can split that jar into four quarters as before. Thus, each of us receives six full jars, one half jar, and one quarter jar. This would be written

$$||| \; \overset{\frown}{||} \; \overset{\frown}{||||}$$

In this way, any amounts—full or partial—can be represented symbolically.

What if seventeen jars of milk need to be divided among six goatherds?

The Egyptian scribes became rather skilled at working with such partial quantities, and a number of papyrus scrolls have survived that were (apparently) used as practice exercises for apprentice scribes in training, such as this one:

By how much does the sum of one-half, one-third, and one-quarter exceed a whole?

This way of handling fractional quantities—breaking them down into increasingly smaller pieces, or *aliquot parts*—has advantages as well as drawbacks, like any other representation scheme. On the one hand, it is a clear and consistent system. When I see the quantity

$$\cap\cap\cap \; \lVert\lVert \; \overset{\frown}{||} \; \overset{\frown}{||||} \; \overset{\frown}{||} \;,$$

I am getting the numerical news in order of importance: thirty-seven whole things (whatever we are counting), plus an additional third, another eighth, and (if I care) a smidgen more, equal to an eleventh part. There's nothing at all ambiguous or confusing about this notation system.

It is, however, a bit long and clunky. If the subdivision gets very fine (yea, verily, even unto the thousandth part), these written and spoken representations could get a little unwieldy.

A somewhat different approach is to subdivide our quantities once and for all, then distribute lots of tiny pieces to each person. For example, when we were sharing the twenty-seven jars of goat milk among the four of us and we each took six jars and had three left over, instead of switching to half jars and then quarter jars we could simply have gone straight to quarter jars and given three to each of us. That is, we would each receive six full jars and three quarter jars.

By modifying the Egyptian notation slightly, and using modern Hindu-Arabic numerals, we can write such quantities very economically, like so:

$6\frac{3}{4}$ "six and three-quarters"

Here the lentil has become a crossbar (or sometimes a slash, as in 3/4), and the 4 underneath indicates that we have chopped our unit into quarters. Taking advantage of the empty space above the crossbar, we then indicate the number of these pieces that we want, namely three.

What would be the modern notation
for  *? How about*  *?*
How would an ancient Egyptian have written $\frac{7}{8}$*?*

This notational device allows us to express all fractional quantities in the same compact and convenient format. For instance, the representation $\frac{15}{32}$ tells us that our unit quantity has been chopped into thirty-two equal pieces and we are taking fifteen of them. Essentially, the bottom number (below the crossbar) tells us what the pieces are (i.e., which aliquot part of the whole), and the top number indicates how many of these we are talking about. The bottom number is known as the *denominator* (Latin for "namer"), and the top number is called the *numerator* ("counter"). This means that the two numbers play different roles, and we'll need to be conscious of that.

Before, when we were discussing multiplication, I was talking about how 3 × 5 does not literally mean the same thing as 5 × 3, since the first number is counting the number of copies of the other. We were fortunate to discover that despite the difference in meaning, the resulting quantities turn out to be the same—multiplication is *symmetric*.

We will have no such luck with fractions. The number $\frac{3}{5}$ is quite different from $\frac{5}{3}$, and not just in appearance. For one thing, the first number is clearly less than 1 (that is, smaller than a whole), because we have chopped our unit into five equal parts (fifths) and taken only three of them, while the second number is grabbing more than three thirds, and so is more than a whole.

Incidentally, it is perfectly meaningful and in no way incorrect to write down a number like $\frac{5}{3}$. In fact, as we shall see, there are definite advantages in doing this, just as there were with "eleventy-seven." Of course, you could also call it $1\frac{2}{3}$ if, perhaps for comparison purposes, you prefer to see how many wholes you've got.

These two representations are the results of two slightly different mental strategies for dividing five by three: we give one to each, then chop up the two leftovers ($1\frac{2}{3}$), or we just chop *everything* into three pieces and give each person five of them ($\frac{5}{3}$).

Which is larger, $\frac{17}{3}$ or $\frac{11}{2}$?

The important thing to understand here is that we are simply *counting*. If I have a number like seven, it means seven of whatever I'm choosing to call *one*; that is, whatever I consider as my unit of thing-ness. It could be seven cows, seven lemons, or, in the case of multiplication, seven sixes or seven hundreds. Similarly, the number $\frac{7}{8}$ is also seven of something—namely, *eighths*.

Just as seven cows added to five cows makes a total of twelve cows, the same goes for counting lemons, sixes, hundreds, or eighths. Thus, $\frac{7}{8} + \frac{5}{8} = \frac{12}{8}$. Much of the art of arithmetic simply comes down to choosing one's units creatively.

In particular, there is a very close connection between fractions and grouping. When we say that seven eighths added to five eighths makes twelve eighths, we are really just adding seven and five. When we go on to say that twelve eighths is the same as one whole and four leftover eighths, it's as if we were members of an octal tribe that groups in eights; we prefer to see twelve organized as one group and four leftovers. Eight ones being a group is the same as eight eighths making one whole.

We can even imagine using Piles of Rocks to assist us in these calculations, where we now include half rocks and quarter rocks and so forth. Or, even more familiarly, we could introduce fractional counting coins as is often done with money: a quarter is one-quarter of a dollar, for instance. Just as we are used to cashing in ten ones for one ten, we can also imagine exchanging four quarters for one whole.

A dozen eggs is both one (dozen) as well as twelve (individual eggs), and the choice of how to think of it is entirely up to you. If you want to think of a half dollar as 2 quarters, 5 dimes, or 50 cents, please be my guest. Every quantity has an infinity of possible names depending on which representation you prefer to work with.

In the modern notation (with numerators and denominators), we can write the same number, say one-half, in a multitude of ways:

$$\frac{1}{2} = \frac{2}{4} = \frac{3}{6} = \frac{4}{8} = \frac{5}{10} = \frac{6}{12} = \cdots$$

These different symbolic representations all refer to the exact same quantity—half of whatever our unit happens to be. If it is a jar of goat's milk, then we are talking about a jar half full (or half empty if you want to be in a bad mood about it).

All we are doing is measuring this quantity using different measuring units. Measured in half jars, it is clearly 1 of them; measured in quarter jars, we have 2; in sixths we have 3— same amount, different grouping sizes.

I suppose, if we wanted to, we could even measure in *thirds* and report that we have one and a half of them. That is, we are perfectly justified in writing

$$\frac{1}{2} = \frac{1\frac{1}{2}}{3},$$

if we so desire. It's a little hard to imagine having that desire, but who knows. Maybe you have a thing for thirds.

How could you write one-half as a number of sevenths?

Truth be told, people don't often chop things up into sevenths or elevenths or other wacky amounts like that. Halves, thirds, and quarters occur fairly regularly—especially in sewing and baking—but I've never seen a recipe that called for $2\frac{3}{5}$ cups of sugar.

The old imperial units tended to use repeated halving, which leads to fractions expressed in halves, quarters, eighths, and sixteenths. The same goes with music, subdivisions of the beat (the rhythmic unit) giving rise to half notes, quarter notes, and so on. Only the most avant-garde twentieth-century composers write scores using eleventh-notes.

As we talked about before, the metric system was introduced to clean up all these messy units and put everything on the same footing—namely, groups of ten—to match the grouping size of the (now universally adopted) Hindu-Arabic system. In particular, this means measuring fractional amounts in tenths, hundredths, thousandths, and so on. The decimal point then marks the division between wholes and parts.

The decimal representation of three and one-half, for instance, would be 3.5, the one-half being measured as five tenths. This means we already have a way to record and calculate fractional amounts—we just express everything in a ten-centric format. Of course, if you are a scientist, or live outside the United States, all of your measurements are already expressed in this convenient way, so there's no need to convert or rethink anything.

On the other hand, maybe you have a one-gram sample of a substance and you need to split it equally among eight test tubes, or you bought five meters of cloth to cut into four matching curtains. How do we represent an eighth of a gram or one and one-quarter meters in a decimal format? An eighth is not a tenth, no matter how hard we squint, just as eight is not ten.

One way to think about fractions is to see them as not only arising from the division process but as being divisions themselves. Thus, the number one-eighth can be regarded literally as one divided by eight. After all, we are chopping our unit into eight equal pieces. This means we can express $\frac{1}{8}$ in decimal terms by dividing the number 1 by the number 8. The clever idea here is to replace 1 by a much larger number, say 1000, and then use place shifting. Essentially, we are regarding 1 as being one thousand somethings (namely, thousandths).

Dividing 1000 by 8 gives us 125, with (thank heaven) no leftovers. (Try it yourself!) This means that one-eighth can also be thought of as 125/1000, or .125 in decimal form. In other words, one-eighth happens to be the same as one tenth plus an additional two hundredths, and then a tiny smidgen of five thousandths more. Thus, any time you find yourself dealing with the number one-eighth, you can instead use the decimal representation .125 if that's more convenient.

What are the decimal representations of $\frac{3}{8}$ and $\frac{5}{8}$?

Similarly, one-quarter can be written .25 and three quarters as .75, which are probably familiar if you live in a place with a reasonable monetary system (unlike Jane Austen and Charles Dickens).

One familiar arena in which fractional quantities are typically expressed in decimal form is in the recording of batting averages in baseball. These are computed by counting the number of times a player gets a hit and comparing it to the total number of opportunities. For example, if Ted Williams was at bat seventy-seven times in a certain season and got a hit (meaning he actually made it on base) twenty-nine times, then his batting average would be the ratio 29/77. In order to compare players with each other (or with past performances by themselves), it is convenient to express all such proportions in the same format, namely decimals, traditionally approximated (or "rounded") to the nearest thousandth. Thus the batting average above would be calculated by dividing 29 by 77:

$$
\begin{array}{r}
376 \\
\hline
77 \overline{\smash{)}29000} \\
231 \\
\hline
590 \\
539 \\
\hline
510 \\
462 \\
\hline
48
\end{array}
$$

Here we would probably round up to .377, since the leftover 48 is more than half of 77. Baseball fans would say, "He batted three-seventy-seven that season" (which, by the way, is sensational).

Another common instance is the use of *percentages*. Here the idea is to express our fractions in terms of hundredths, thus measuring a given ratio as so many out of a hundred (or *per centum*, whence the abbreviation "percent"). Thus, a fraction like $\frac{1}{4}$ could be expressed as 25 percent, meaning 25/100. Of course this is nothing other than .25 dressed up in slightly different clothes.

There is an amusing notational history here, by the way. Just as medieval scribes got tired of writing *et* (the Latin word for "and") every five seconds and began abbreviating it using the ampersand symbol (&), the notation "per 100" was gradually shortened and smeared together to form the *percent sign* (%). Thus, the number 25/100, or twenty-five percent, became 25%. Essentially, the percent sign is just shorthand for hundredths.

All that's going on here is that we are adapting our various fractions (with whatever denominators they may have) to a base-ten culture, just like with the metric system. All that is required is to do some division and then ignore the leftovers once they become insignificantly small.

Of course, if the fractional quantity you are interested in *already* has a denominator of ten or one hundred, then we don't need to do any converting at all: .01 already means one-hundredth. So a number like $\frac{37}{100}$ (or 37%) would be rendered in decimal form as .37, or 37 with two antishifts, if you like.

On the other hand, ten is a perfectly arbitrary grouping size, and not every fraction is going to get along with it so nicely. For example, if we try to express the number $\frac{2}{3}$ in decimal by dividing 2000 by 3, we encounter this amusing phenomenon:

$$\begin{array}{r} 666 \\ \hline 2000 \\ 18 \\ \hline 20 \\ 18 \\ \hline 20 \\ 18 \\ \hline \end{array}$$

Each time we give away three sixes (in whatever place), we get a remainder of two, and we enter an infinite loop. This is both annoying and really cool. So $\frac{2}{3}$ is not any nice number of tenths or hundredths or thousandths—it always leaves a remainder no matter how far we go.

What we're seeing is that three and ten simply do not get along. Thirds are not very conducive to being represented by tenths. Similarly, if you like to group in threes and you build a whole place-value system around that grouping size, tenths would not be very cooperative.

So not every fractional amount can be expressed precisely in decimal format. That is, unless you want to allow infinitely long representations. You can, in fact, express the number $\frac{2}{3}$ in the form .666... (where the sixes continue indefinitely). This is an interesting idea in principle but clearly quite impractical. (The helpful smile will quickly fade from the draper's face as you take out your electron microscope to measure another six nanometers of chintz.)

Of course, in real life nobody measures anything that precisely, for the simple reason that we can't. Nor would it make any sense to do so. How accurate do a carpenter's measurements need to be? Wood expands in the summer and absorbs water when it rains. Besides, if a piece of wood is a few hundredths of an inch too long, then sand it down. That's what sandpaper and tile grout are for. Exactitude in carpentry, sewing, and baking is preposterous—estimation and approximate rules of thumb are far more valuable.

This is true for scientists and engineers as well, though the tolerances are somewhat more precise. Whereas a carpenter or a seamstress would be happy to ignore a millimeter here or there, a nuclear physicist may require an accuracy of seven or eight additional decimal places. Still, when it comes to a billionth of a billionth of a millimeter, we simply lack the equipment to make such fine distinctions.

Moreover, the physical universe itself seems to have a few things to say about our ability to measure it precisely. Not only is exact measurement unnecessary and impractical, it is also (by the apparent nature of physical reality) *impossible*. How long is my nose, exactly? Where exactly does nose end and air begin? In fact, molecules of air and nose are trading places and jiggling around constantly. Noses are only statistical, so they don't really have a precise length. And neither does anything else in this universe.

Of course, there are other universes. We can imagine a perfect world of exact measurements and see where that leads us. Mind you, this would be a purely philosophical investigation, not so much motivated by any practical concerns but for the sheer intellectual pleasure and entertainment it might provide. Also, despite the fact that the world we live in is fuzzy, random, and inexact, it's often pretty close to exact, and a perfect mathematical realm (which would necessarily be fictitious) may give us a new perspective on reality and also a new understanding and appreciation for numbers and their properties in the abstract.

Not to mention how much simpler and more elegant such a place would be. This place, which I like to call Mathematical Reality, is actually where I spend the bulk of my time, as do most of us mathematicians.

All right, so what if we accept such a reality. Then $\frac{2}{3}$, for instance, is an exact quantity and is decidedly not equal to .6666 or any other approximation. Two-thirds is exactly what it claims to be: two of those things called thirds. And a third is simply the thing that, were you to put three of them together, would make the number one.

But one *what*? What are we counting and measuring in this purely mathematical realm? This is one of the first great examples of the simplicity and abstract beauty of mathematics. We aren't talking about cows or lemons at all. When we say "one" we don't mean one actual real-life object, we mean one in the abstract—the imaginary entity that holds the information of *oneness*. (Do I sound like a New Age guru?)

So along with the usual "counting numbers" one, two, three, and so on, we now have a whole slew of additional numbers like one-half, two-thirds, fifty-seven-elevenths, and so on. We know exactly what we mean by these things, and we have a perfectly nice representation system for them—namely, the numerator and denominator language: $\frac{1}{2}, \frac{2}{3}, \frac{57}{11}$. There are no perception problems here; we know at a glance exactly what we've got. The notation $\frac{57}{11}$ precisely indicates the number obtained by chopping one into eleven equal parts and taking fifty-seven of them.

With new numbers to talk about, we have a host of new arithmetic problems—new opportunities to exercise our wit and imagination. The most immediate question is, how do we compare such quantities?

Of course, if the two fractions we wish to compare happen to have the same denominator, then it's easy: $\frac{9}{11}$ is larger than $\frac{7}{11}$ because they are both counting elevenths, and nine is bigger than seven.

Similarly, the comparison is also straightforward if the two numbers have the same numerator. Now both fractions are counting the same number of pieces, but the pieces are not the same size. For instance, we can easily tell that $\frac{4}{5}$ is larger than $\frac{4}{7}$, because a fifth is bigger than a seventh. The more pieces you chop something into, the smaller the pieces must be.

Which is larger, $\frac{3}{7}$ or $\frac{2}{5}$?

The difficulty here lies in the fact that we are using different size pieces. One number is counting sevenths and the other is talking about fifths—we're comparing apples and oranges.

This is actually a familiar problem. The same thing happened with multiplication, you may remember. The trouble with comparing 5 × 8 and 6 × 7 is that one number is currently grouped in eights and the other is a bunch of sevens. So what did we do? We rearranged them for ease of comparison—that is, we did some arithmetic.

More precisely, we converted both representations to the same grouping size. We could rearrange five eights into groups of seven, or put the six sevens into piles of eight, or, if we preferred, we could convert both amounts into some third convenient grouping size (typically tens). The important thing is to get a common language so we can more easily understand what's going on.

The same thing is going on here with fractions. In a sense, a number like $\frac{3}{7}$ is really a multiplication: it is three copies of a group, where the grouping size is *sevenths*. So, because one of our numbers is grouped in sevenths and the other in fifths, what we need is a common grouping size that works for both—one that allows us to express both of our quantities easily and precisely.

Tenths, for instance, would not be advisable here. Although the number $\frac{2}{5}$ is amenable to being expressed nicely in tenths (it would be four of them), the number $\frac{3}{7}$ is not so happy. Of course we could approximate it using a decimal representation and then use that to compare:

$$\frac{2}{5} = .400...$$
$$\frac{3}{7} = .428...$$

This does indeed show us that $\frac{3}{7}$ is slightly larger than $\frac{2}{5}$, though it requires a bit of work to produce the approximations. (I would suggest using a calculator.) Also, it's kind of ugly and technical feeling.

And there is a subtle problem with this approach. What if two fractions are really, really close but not equal? Then their decimal representations will agree in the first several decimal places, and you'll have to keep dividing until you encounter a mismatch. Only then will you know which one is larger. Even worse, if the two fractions happened to be exactly equal, but not obviously so due to their different representations (e.g., $\frac{91}{117}$ and $\frac{273}{351}$), you would continue dividing and producing decimal digit expansions that always agree, no matter how far you carry out the process. Then you would never know if one of them were larger than the other—the digit that reveals this fact might be just around the corner. This is way too cumbersome and time consuming an approach.

Better would be to find a "common denominator"—a grouping size compatible with both fifths and sevenths simultaneously. Then both numbers could be expressed in the same language. Essentially, we will be *renaming* our fractions to make comparison easy and straightforward.

We saw before that every fraction can be represented by a numerator and a denominator in infinitely many ways. Thus, the number $\frac{1}{3}$ is the same as $\frac{2}{6}$, which is also equal to $\frac{3}{9}$. This is because any subdivision of a unit into smaller pieces can always be further subdivided if we wish.

Just as with ordinary counting, if I am keeping track of egg cartons (i.e., dozens) and you are counting individual eggs, our numbers are going to differ because our units do. Similarly, if I am counting in thirds and you are counting in sixths, then you will end up with twice as many pieces as I. This is simply because each of my pieces is worth two of yours.

The point is that if your units are smaller than mine, then you will need more of them to measure the same quantity. If my unit is exactly five times larger than yours, then you will need five times as many of yours as I need of mine. Make good sense?

The pattern here is that a fraction is unchanged if both its numerator and denominator are enlarged by the same factor—that is, multiplied by the same amount. For example, $\frac{2}{3}$ can be rewritten as $\frac{8}{12}$. Here we have multiplied (or "scaled") both the numerator and denominator by a factor of four. So instead of measuring in thirds and counting two of them, we are chopping our unit into four pieces to make twelfths, and consequently we need four times as many of them, namely eight.

Thus we are free to "blow up" both the numerator and denominator of a fraction by the same factor without changing the number itself. This gives us some flexibility in choosing our names for things, just as we are free to change grouping sizes (which, in fact, is exactly what we are doing, if you think about it).

Can the number $\frac{49}{21}$ be written with a smaller numerator and denominator?

Speaking of blowing up, one way to view a fraction like $\frac{3}{7}$ is to imagine a "number enlarger" that allows you to blow up (or shrink down) numbers, just as a copier does to printed images. You can set the enlarger to double or triple the size of a number, and you can also shrink it down by a given factor. Then we can think of $\frac{3}{7}$ as the result of putting in the number 1, shrinking it down by 7 (i.e., making it one-seventh the size), and then blowing it up by 3 (that is, tripling it).

Conveniently, the results of these processes are independent of the order in which they are performed. If you blow up by 2 and then by 3, you have effectively blown up by a factor of 6, which is also what you would get if you tripled first and then did the doubling. In the same way, we could also get $\frac{3}{7}$ from 1 by tripling first (to get 3) and then shrinking down by 7. This is because 3 ones, divided by 7, is the same as 3 "shrunken ones," each of which is one-seventh of the original. It's the same thing as before when, to multiply some pennies and dimes by ten, we imagined the pennies transforming into dimes and the dimes into dollars.

What we're saying is that at any time we wish, we can freely rename or regroup a given fraction by blowing up (or shrinking down) both the size of the pieces (the denominator) along with the number of them (the numerator).

So how does this help us compare two fractions? The idea is to use scaling to rewrite both numbers so that they share the *same* denominator. Then they will be counting the same size pieces, so then the numerators will tell us the story. For example, the numbers we had before, $\frac{3}{7}$ and $\frac{2}{5}$, are hard to compare directly because 5 and 7 are different. Let's list some alternative names for them:

$$\frac{3}{7} = \frac{6}{14} = \frac{9}{21} = \frac{12}{28} = \frac{15}{35}$$

$$\frac{2}{5} = \frac{4}{10} = \frac{6}{15} = \frac{8}{20} = \frac{10}{25} = \frac{12}{35} = \frac{14}{35}$$

Aha! We discover that among the various representations of these two numbers, there is a common denominator, thirty-five. So hiding beneath their mild-mannered appearances, $\frac{3}{7}$ and $\frac{2}{5}$ have "secret identities" $\frac{15}{35}$ and $\frac{14}{35}$, respectively.

This makes it laughably easy to compare the two: $\frac{3}{7}$ is a little larger than $\frac{2}{5}$. In fact, it's also easy to see exactly how much larger, namely $\frac{1}{35}$. So the common denominator idea is really the nicest approach. Don't change anything, don't bother approximating, just rewrite and rename until it becomes obvious.

Of course, this raises the question of how best to go about finding a common denominator in the first place. We got ours by trial and error—listing each possibility until we found denominators that matched. That's pretty slow and tedious. There's actually a much easier (and decidedly more clever) way.

Notice that we began with denominators 7 and 5 and ended up using a common denominator of 35, which happens to be the product of the two. This is no coincidence. In fact, the easiest way to get a common denominator is to simply blow up each fraction by the denominator of the other—that is, we scale the numerator and denominator of $\frac{3}{7}$ by 5, and we scale the top and bottom of $\frac{2}{5}$ by 7. Since the new denominators will now both be equal to the product of the old ones, they will necessarily be the same. Each number simply looks to the other for its choice of scaling factor.

Let's try it out. Suppose we want to know which is larger,

$\frac{3}{8}$ or $\frac{2}{5}$. (We just saw that $\frac{2}{5}$ is smaller than $\frac{3}{7}$, but $\frac{3}{8}$ is also slightly smaller than $\frac{3}{7}$, so it's not clear.) All right, let's use our new method: we rescale $\frac{3}{8}$ by 5 and $\frac{2}{5}$ by 8 to get

$$\frac{3}{8} = \frac{15}{40} \text{ and } \frac{2}{5} = \frac{16}{40}.$$

Now it is easy to see that $\frac{2}{5}$ is the larger one. And we get the free bonus of knowing they are off by exactly $\frac{1}{40}$, should that be of interest.

Thus we see that for the purposes of addition, subtraction, and comparison, having the same denominator makes life much simpler. All quantities are measured with the same unit; all the pieces have the same name. For example, the sum of these two numbers, $\frac{3}{8} + \frac{2}{5}$, is easy to calculate in the rewritten form

$$\frac{15}{40} + \frac{16}{40} = \frac{31}{40}.$$

Now we can see that the total is just a smidgen (namely $\frac{1}{40}$) larger than $\frac{3}{4}$.

Is $\frac{1}{2} + \frac{1}{3} + \frac{1}{5}$ larger than 1?

As I mentioned before, there really is no compelling rationale for making such calculations in the real world. Using a standard ten-digit pocket calculator, we could simply have the machine perform the requisite divisions and so forth to obtain accurate approximations that are more than adequate for any practical workaday purposes. If for some reason (and it's a little hard to imagine such a scenario) you really needed to calculate $\frac{2}{7} + \frac{5}{8}$ to see if it was safely below 1, you could simply pick up the nearest phone or wristwatch and punch in 2 ÷ 7 + 5 ÷ 8 = to get 0.910714285, and your problem is solved.

Oh, that reminds me. I forgot to mention the division symbol (÷) that often appears on calculators and covers of math textbooks. Of course, it's just the fraction crossbar with dots indicating where the numbers would go. Thus people write 12 ÷ 3 = 4, and so forth.

So the arithmetic of fractions is not really of any practical concern. It's more about numbers and their patterns of behavior. If that is of no interest to you, then we're done here, and you can close this book right now. On the other hand, if you want to see something elegant and intriguing, albeit abstract and useless, then please read on!

What is the sum of $\frac{2}{7}$ and $\frac{5}{8}$ exactly?
How far off from 1 is it?

Using the common denominator approach, we reduce addition, subtraction, and comparison of fractions to working with numerators only—that is, we're back to simple whole-number counting.

But what about multiplication and division? Can we have $2\frac{1}{2}$ copies of $3\frac{1}{4}$? What about $\frac{3}{8}$ divided by $\frac{4}{5}$? Do such questions even make sense? Who needs to share $17\frac{1}{2}$ jelly beans among $4\frac{1}{2}$ people?

Before we delve into these questions, I want to say a few words about the various representation options we've discussed, along with their advantages and disadvantages. This is mostly an issue of aesthetics, so I'll really just be offering up my own personal tastes and opinions (which, as it happens, I'm not terribly averse to doing).

Let's imagine we have a certain fractional quantity, say two and one-quarter. Among the many options (forgetting about Egyptians or Banana tribesmen for the moment), we have the standard Hindu-Arabic representations

$$2\frac{1}{4}, 2.25, \frac{9}{4}.$$

The first version is what you typically see in cookbooks and carpentry manuals. The advantage is that it makes the comparison news very clear (it's a bit larger than two), and is a good choice when you are working with natural physical units (e.g., rulers and measuring cups) and you want to get the whole units out of the way first.

The second form is preferred by scientists and engineers who deal in highly accurate approximations and want a standardized scale for comparison. Hence the metric system and all that.

You may sneer and turn up your nose at the third option, $\frac{9}{4}$, as being top heavy and impractical. After all, it's not even obvious how big it is. We would have to do some reorganization (that is, arithmetic) to determine if it is larger than two, for instance.

On the other hand, it is the only form I would even consider using for mathematical (that is, purely theoretical) purposes. All the information is there in compact form. I can easily see it as being nine of something (namely quarters), and it will be much simpler to operate on as a result. Moreover—and this is the real advantage—it puts all of our numbers in the same linguistic category and makes no distinction between wholes and parts. In this context, I might even prefer to write a number like six as $\frac{6}{1}$, as ludicrous as that might seem.

How would you write eight and one-third
in these three formats?

My advice (taste, really) is to use wholes and parts (e.g., $2\frac{1}{4}$) when sewing and baking cookies, decimal expansions (2.250) in the laboratory, and pure ratios ($\frac{9}{4}$) when investigating numbers and their properties. Or do whatever you want whenever you want (which is my real advice, anyway).

One of the things that makes thinking about fractions fun and enjoyable, as well as potentially confusing, is the multitude of possible interpretations. The number $\frac{2}{3}$ can be regarded as a *measurement* (breaking our unit into three equal parts and taking two of them), a *division* (the result of sharing two jelly beans among three people), a *ratio* (holding the information that a pair of quantities are in proportion as two is to three), a *scaling* (blowing up by 2 and shrinking down by 3), or, most abstractly, as simply being "the entity that when multiplied by three is two." Each of these viewpoints carries with it a set of mental images and connotations, which can be useful in different circumstances.

When it comes to multiplication and division, for instance, I often like to think in terms of scaling. Multiplication by two (i.e., doubling) is blowing up by a factor of two. Division by two (that is, cutting in half) can be seen (and *felt*) as shrinking down by two. This does not mean taking two away but rather scaling it down to half its former size. Two is the scaling factor by which the entire quantity is being shrunk or enlarged, as if it is being fed into some sort of abstract numerical copy machine.

Multiplication is often expressed verbally (at least in English) using the preposition *of*—"I'd like eight of those yummy muffins, please" or "Hey, someone ate two-thirds of my yummy muffin!" To multiply a quantity by a fraction is to specify that much of it, just as multiplying an amount by three is to take three of it.

So what we would mean by one and one-half times six is that we want one and a half copies of six—that is, one whole copy of six and then an additional half of a six, which would be three. Thus, $1\frac{1}{2} \times 6 = 9$.

Alternatively, we could think of $1\frac{1}{2}$ as being $\frac{3}{2}$, and that we are interested in "three halves of six," which, just as it says, means three "halves of six," that is, three threes, or nine, as before. Or, if you prefer, we can imagine that we are putting 6 into the copy machine and asking it to shrink it down by 2 and then blow it up by 3. We could just as well reverse the order and have it blow up by 3 first (to get 18) and then shrink down by 2 to get 9. I hope this all makes sense to you and that you find it amusing and enjoyable to have so many cute alternative ways to think about it.

Another one just occurred to me, actually. Using the symmetry of multiplication, one and a half sixes could also be treated as six copies of $1\frac{1}{2}$. Since two copies would make 3, and six is three twos, again we see that it's the same as three threes.

In general, I think the copy machine metaphor works best. Suppose we wanted $3\frac{1}{2}$ copies of $2\frac{1}{3}$, for instance. This kind of thing can actually occur in real life (e.g., when scaling a recipe). There are many ways to proceed here, but the way I usually like to operate would be to first express everything in purely fractional terms: $\frac{7}{2} \times \frac{7}{3}$. Then I can view each of these numbers as a set of copy machine instructions, that is, I am sticking the number 1 into the copier and asking it to blow up by 7 and shrink down by 3. That gives me my $\frac{7}{3}$. Then I proceed to blow up again by 7 and down by 2. That multiplies it by $\frac{7}{2}$. All told (and in any order I please) I have started with 1 (of whatever my unit is, if I even have one), done two blow-ups by a factor of 7 and two shrink-downs, by factors of 2 and 3. This can be rephrased as a blowup by 49 and a shrink-down by 6. In other words, I'm left with the number $\frac{49}{6}$. (And the copier is now making an annoying beeping noise because I left my original on the glass.)

So we find that $\frac{7}{2} \times \frac{7}{3} = \frac{49}{6}$. Of course, if you want to rewrite this quantity as $8\frac{1}{6}$, be my guest. If you work in a bakery and need to measure out that many cups of flour, then by all means do—eight full cups and an additional one-sixth of a cup is a lot less scooping and pouring than doing forty-nine individual sixths of a cup, that's for sure. But if you are a mathematician interested in numbers for their own sake, then $\frac{49}{6}$ might be a more convenient and informative representation. For one thing, it reveals clearly the behavior of fractions under multiplication: to multiply two fractions we can simply multiply their numerators and denominators separately. This is what the copy machine interpretation shows us. So that's very nice, and that's why I like to work this way.

Which is larger, $3\frac{1}{2}$ copies of $2\frac{1}{3}$,
or $2\frac{1}{2}$ copies of $3\frac{1}{3}$?

Division of fractions does not occur terribly often in everyday life, although one could imagine an occasional scenario such as the following: A large vat containing 200 gallons of maple syrup needs to be poured into individual jugs, each of which holds $1\frac{1}{3}$ gallons. How many jugs will be required?

Essentially, we are asking what number, when multiplied by $1\frac{1}{3}$, makes 200. In other words, what is 200 divided by $1\frac{1}{3}$. Weirdly, it's the same as if we were trying to share 200 jelly beans among $1\frac{1}{3}$ people. The symmetry of multiplication means we can entertain either view.

Before we analyze this situation further, you may have noticed something about multiplication and division—in a way, they're sort of the same. Dividing by two, for instance, is exactly the same thing as multiplying by one-half. What this means is that we don't really need a shrink-down button on our copier; we could just blow up by one-half or one-third or whatever. The distinction between enlarging and reducing, blowing up or shrinking down, is merely the difference between multiplying by numbers larger or smaller than 1. Maybe humans care about that distinction, but numbers don't seem to. We multiply by two or we multiply by one-half. Either way, we multiply.

Of course, I'm not trying to say that doubling and halving are the same process, or that 2 and $\frac{1}{2}$ are the same number. But there is a close connection between them: dividing by one is the same as multiplying by the other. This is what is known as a *reciprocal relationship* (from Latin *reciprocare*, "to go back and forth"). The numbers 2 and $\frac{1}{2}$ are said to be reciprocals of each other.

What this means, really, is that they have opposite effects under multiplication. One way to undo doubling is to multiply by one-half. Another way to say it is that 2 and $\frac{1}{2}$ cancel each other out—if you blow up by 2 and then blow up by $\frac{1}{2}$, you end up back where you started. This is because $2 \times \frac{1}{2} = 1$, and blowing up by a factor of 1 is the same as doing nothing. So two numbers being reciprocals of each other means their product is 1. The reciprocal of a number is its multiplication partner—the thing that it multiplies with to make 1. Thus $\frac{1}{2}$ is the reciprocal of 2, and 2 is the reciprocal of $\frac{1}{2}$.

What is the reciprocal of $\frac{2}{3}$?

When we are dividing, we are asking an implicit question: What do we have to multiply by to get the total we want? Calculating $100 \div 7$ is really just asking what number of sevens makes a hundred. Or, perhaps more suggestively, how do I "unmultiply" 100 by 7?

If we think of multiplication by a number as a process that we can perform on quantities, then division can be seen as the reverse process. To undo multiplication by 7, we divide by 7. Or, as we've seen, we can multiply by $\frac{1}{7}$. This means we get to make an interesting trade: instead of undoing the multiplication activity with division (i.e., the inverse activity), we undo multiplication with another multiplication! That is, we are transferring the undoing responsibilities away from the operations and over to the numbers themselves. Instead of unmultiplying by seven, we are multiplying by *unseven*. Rather than putting on my shoe and then taking it off (which requires two different operations, putting on and taking off), I can imagine that all I ever do is put things on. In the morning I put on my shoes and in the evening I put on my "antishoes," that is, my reciprocal shoes that cancel and annihilate my regular shoes. I make my shoes do the work so I don't have to.

So what does it mean to divide a number by $\frac{1}{2}$, say? It does not mean to divide it in half—that's what dividing by *two* does. We're dividing *by* one-half, not *in* half. If we want 5 divided by $\frac{1}{2}$, we are asking how many copies of $\frac{1}{2}$ make 5, or what do I have to cut in half to get five? Clearly, it's ten. So 5 divided by $\frac{1}{2}$ is 10? We divided a number and it got bigger. Does that make sense?

Perhaps we are used to dividing by numbers larger than 1, which means we would be shrinking down. Here, however, we are dividing by a small number, so it takes more copies of it to make the required total. Specifically, it takes ten halves to make five wholes.

Here we see the beauty and symmetry of the reciprocal relationship. Just as dividing by 2 is the same thing as multiplying by $\frac{1}{2}$, we find that division by $\frac{1}{2}$ has the exact same effect as multiplying by 2.

Let's see if we can understand this behavior more deeply. We are choosing to view division as a kind of unmultiplication. If we imagine our copier as having an input slot and an output slot, so that when we stick in a number it comes out seven times larger, then division by seven could be thought of as taking a number and trying to shove it into the output slot, running the machine in reverse. (Don't try this with a real copier!) What I'm saying is that this would be the same as running it through the machine normally but with the scaling factor set to $\frac{1}{7}$ instead of 7. Again, it's just what I was saying about trading inverse processes for inverse numbers.

Suppose now that I have the machine set to multiply by $\frac{2}{3}$. As a process, this means that I am blowing up by two and shrinking down by three. Clearly the inverse (or backward) version of this would be to blow up by 3 and shrink down by 2. That is, to undo multiplication by $\frac{2}{3}$, you just need to multiply by $\frac{3}{2}$. The reciprocal of a fraction simply interchanges numerator and denominator. How beautifully unexpected and yet so perfectly sensible.

There is one amusing exception to this pattern, actually, and that is the number *zero*. Not that anyone would ever have reason to divide a quantity into no parts at all (whatever that might possibly mean). My point is that if we are going to reimagine division as multiplication by the reciprocal (or running the copier in reverse), we should be aware that the metaphor crashes to the ground in the case of multiplication and division by zero. The reason is that multiplication by zero destroys information. Any number times zero is still zero, or to put it another way, blowing up by a factor of zero reduces everything to a single point. No matter what number you put in the copier, zero keeps coming out. This means there cannot be a reverse process. Some activities are so destructive they cannot be undone. Still another way to say it is that since 0 times anything is always 0, in particular it can never be 1. So zero (and zero alone) fails to have a reciprocal. Other than that small proviso, everything works fine.

So division by fractions (and that includes whole numbers if you want to think of them as having a denominator of 1) is the same as multiplication by their reciprocals, and these can be easily obtained visually by flipping them upside down, as it were. That's both pretty as well as convenient.

It is also *dangerous*, like many things in mathematics that obey simple, elegant patterns. The danger is that people come to think of the visual and tactile symbolic manipulations as definitional rather than behavioral. The number 5 × 3 does not *mean* 3 × 5; the two numbers happen to be equal, and that is both beautiful and fortunate, but it is not built into the definition of multiplication as making copies—rather, it is a *discovery* that demands an *explanation* (which we in fact have, in the form of rows and columns of rocks, if we like).

Similarly, division does not mean "flip and multiply" (as some unfortunate victims of compulsory schooling seem to believe), any more than taking off my shoe means putting on my antishoe. That's not what we mean by *meaning*. That's what we mean by viewing, considering, deducing, or interpreting. Numbers aren't symbols on a page or screen, and arithmetic operations are not patterns of motion or transmutation of these symbols. However, it certainly is wonderful that we can encode everything in such a way that the notation is easy to work with, and I'm certainly not knockin' it. I just like to be clear in my own mind as to what's what.

To return (finally) to our maple syrup example, we wanted to divide 200 gallons of syrup into $1\frac{1}{3}$-gallon jugs. The number of jugs would then be 200 divided by $\frac{4}{3}$. If we like, we can view this as $200 \times \frac{3}{4}$, or $\frac{600}{4}$. So that's how many jugs we need. Of course, we're not really interested in quarter jugs; we want to measure in whole jugs (and not jugs with holes, we hope).

One easy way to divide by four (that is, to quarter some-thing) is to cut it in half twice. So 600 becomes 300, then 150. Perfect! No leftovers to worry about. So we can empty our 200-gallon vat into exactly 150 jugs.

What would have happened if we had 201 gallons?

To sum up, here's the deal with fractions as I see it. First off, they are in no way necessary in the practical everyday world. Even if you live in a backward country like the United States that still uses outdated imperial units, you can still just take out a calculator and do all of the necessary computations. Put everything in the same decimal format, and let the machine do its thing. Then use as many digits of this approximation as are appropriate.

Of course, there's more to life than just getting things done; there's learning and understanding, joy, love, and fun to be had. And if you are intrigued by numbers and other abstract imaginary constructions of the human mind, then there's mathematics. Numbers have beautiful and amusing properties for us to discover and investigate, and fractions are an espe-cially elegant and interesting mathematical environment.

They are particularly good at holding onto exact measure-ment information (which is why they are particularly unfit for practical and scientific purposes). Any quantity that can be obtained from a unit by repeated chopping and copying can be held in the form of a pure fraction with a numerator and a denominator. We can then add, subtract, multiply, divide, and compare such quantities easily.

The simplest pattern is with multiplication—the numera-tors and denominators just get multiplied independently (e.g., $\frac{2}{3} \times \frac{5}{7} = \frac{10}{21}$). Division is the same; we just view it as multiplica-tion by the reciprocal, for example, $\frac{2}{3} \div \frac{5}{7} = \frac{14}{15}$.

For addition, subtraction, and comparison, we usually want to express everything in the same grouping size, so that means scaling to obtain common denominators. Then we can simply add, subtract, or compare the corresponding numerators. For example, $\frac{3}{4}$ and $\frac{2}{3}$ can be dealt with easily, once we rewrite them as $\frac{9}{12}$ and $\frac{8}{12}$, respectively. In this way, we see that we are really just dealing with the numbers 8 and 9; the fact that they are counting twelfths can be pushed to the background. The arithmetic of fractions is really just the arithmetic of whole numbers (i.e., counting), as long as you are clearheaded about your units.

Which is larger, $\frac{3}{5} \times (\frac{2}{7} + \frac{1}{3})$ or $\frac{13}{11} \times (\frac{3}{5} - \frac{2}{7})$?

NEGATIVE NUMBERS

As useless and unnecessary as they may be from a practical standpoint, these new numbers—the perfect fractions—are certainly simple and convenient and possess many beautiful and elegant properties. And the fact that they can be used to approximate real-world measurements to arbitrarily high precision allows us to construct simplified models of reality that are easier to think about and work with. A real desktop is a nightmarish tangle of organic compounds—wood fibers, water, oils, not to mention all the dust mites and bacteria crawling all over the place. When we replace this fuzzy, hairy, jiggling mass of atoms with an imaginary perfect rectangle, we dispense with all of this complexity and move to a quieter, more peaceful realm of simplicity and exactitude.

Of course, the extent to which this is a good idea will depend on what you are doing and what you care about. If you are planning to paint and decorate this desktop, and its corners happen to be quite rounded, you may need to incorporate this fact in your mathematical model. We all want to simplify, but sometimes we can go too far and leave out important information.

The same goes with measuring. If we get out a tape measure and find that the desktop measures (approximately) $32\frac{1}{2} \times 18\frac{3}{16}$ inches, then calling it a 32×18 rectangle might not be good enough. Depending on what we're doing, the ignored half-inch might actually matter. On the other hand, there is *some* level of precision (perhaps extremely fine) after which we no longer care, or can no longer tell the difference. Let's suppose that in our case we're happy to ignore things smaller than an eighth of an inch, say. Then we can round off the $18\frac{3}{16}$ to a simpler $18\frac{1}{4}$, and model our desktop with a mathematical rectangle of exact dimensions $32\frac{1}{2} \times 18\frac{1}{4}$. If we wanted, we could even make a scale model using a quarter-inch as our unit. Then, rewriting these fractions as $\frac{130}{4}$ and

$\frac{73}{4}$, respectively, we can imagine our abstract rectangle to have whole number dimensions 130 × 73 units. We could then proceed to plan our painting and decorating project, away from the actual wooden object, using this more portable and convenient mental construct.

This is pretty much what scientists and engineers spend their time doing. They construct simplified mathematical models of their problems and then work on them using the precise and elegant patterns that mathematical objects obey. Then they translate back to reality, making any necessary approximations as they see fit. In this way, mathematics can be viewed as a "model shop"—a source of convenient simplified models of reality.

For mathematicians, however, the correspondence works the other way. The objects of interest *are* the abstract imaginary patterns, and reality serves merely as a source of modeling material. If I want to think about perfect imaginary rectangles, I can use a desktop or a piece of paper as a crude, admittedly inaccurate physical model. Of course, we understand that such a clumsy prosaic object could never contain any actual mathematical truths (in particular, it cannot have exact measurements), but it can still give me ideas and lead me to observations and understanding nonetheless. So for mathematicians like myself, reality is the model shop. This was more or less Plato's view: real things are merely the shadows of the true, idealized mathematical forms.

In particular, numbers themselves have such a dual existence. When you say that you have five lemons in your basket, or that you are five feet tall, we all understand (at least implicitly) that this is a crude, real-world statement. Questions of exactly what is meant by "lemon" or "foot" are not generally asked. Without really giving it much thought, we tend to model the situation mathematically by ignoring the feet and lemons and simply encoding the information using the number five.

But what really is *five*, in the abstract? Five what? I suppose we could say five *ones*, but then we have to address the question of what "one" means mathematically. What exactly is unity, divorced from any reference to reality?

Ignoring the long (and to my mind rather pedantic and trivial) history of Greek and medieval scholastic thought on the subject, I'd like to tell you about the way modern mathematicians think about numbers, or at any rate how I like to think about them.

I think I can sum up the modern mathematical viewpoint in a single phrase, which we could take as the algebraists' creed: *a number is what a number does.* What we're saying is that it doesn't really matter what numbers are, it matters how they act, or behave; that their behavior is tantamount to a defining characteristic, or implicit description. Does it really matter what a duck is, as long as it swims and quacks?

Mathematics is the study of *pattern.* And of course I mean pattern in the abstract—the patterns of Mathematical Reality. Mathematical structures, such as triangles and numbers, enjoy a wide variety of intriguing and amusing properties, and it becomes almost impossible not to endow them with a kind of existence. They become "creatures" of Mathematical Reality, and the various patterns we discover are their "observed behavior."

In particular, numbers start to feel less like quantities and more like *entities*—creatures that interact with each other and engage in curious and intricate arithmetical dances.

Thus, the number $\frac{2}{3}$, rather than representing two-thirds of some quantity, such as a chocolate bar or a meter stick, can be viewed abstractly as "that which when multiplied by 3 is 2." As an imaginary mathematical entity, it is completely defined and determined by what it does. Anything with the property that when multiplied by three is two serves as an embodiment of two-thirds-ness.

OK, you might say, but what are two and three themselves then; what is it that they do? The modern view would be that whole numbers like two and three are simply the result of addition: 3 is shorthand for 2 + 1, and 2 itself is merely an abbreviation for 1 + 1.

But what is 1? Surely the buck has to stop somewhere, doesn't it? We can't just keep defining everything in terms of everything else. Or can we?

It turns out that we can, actually. The number zero, for example, can be defined behaviorally as well. Rather than a symbol representing nothing, or the number of lemons you have when you haven't any lemons at all, we can instead think of the number zero as being the entity with the very specific property that when added to numbers leaves them alone.

Similarly, we can define the number 1 to be the thing that when multiplied by a number leaves it unchanged. Here we see these two very special creatures, zero and one, being characterized not by their real-world images (i.e., as sizes of small collections) but operationally, in terms of their behavior and relationship to other numerical entities. Instead of counting nothing or being nothing, we're seeing zero as *doing* nothing (at least with respect to the addition operation).

But what is adding, then? Isn't it pushing piles of rocks together? If we're thinking of numbers not as counting or measuring but as "behaving" in some abstract sense, then what is the viewpoint on operations like addition and multiplication? How can there be adding without counting?

Again, the modern idea is to forget where the concrete real-world notion came from and to focus our attention on the behavioral patterns. The question becomes not what addition is, or what it means, but what patterns does it exhibit. What are the outward behaviors that characterize addition, as opposed to some other activity?

It's high time for an analogy of some sort. Let's suppose you have a cage full of hamsters and you are interested in their behavior. You may notice that some of them engage in various mating rituals and other social behavior, and you may also have observed that some of the hamsters appear to have special status within the hamster community. There are both general properties shared by all hamsters (e.g., they all have a skeleton), as well as particularities that make each individual hamster unique and special. This is more or less the way a modern mathematician (this one, anyway) feels about numbers.

In particular, my imaginary hamsters like to do things that normal hamsters cannot: they like to "combine" to form other hamsters. We could attempt to shoehorn in some kind of reproductive metaphor, but it's not really very apt. The thing about mathematical operations (e.g., addition and multiplication) is that they don't really "create" new numbers; they interrelate the ones we already have. The simplest way to think about operations is not as an activity or a process like combining or joining but more abstractly as an *assignment*. That is, to each pair of hamsters we simply designate a certain hamster as their "sum." All arithmetic operations are then on the same equal footing as being abstract assignments of numbers to pairs of numbers. The question then becomes how these various assignment patterns differ from one another.

So the question we (as mathematicians) would then ask is, what does addition do? How does it distinguish itself from a purely random assignment of hamsters to pairs of hamsters? Well, for one thing, addition is *symmetrical*. That is, it does not depend on the order of the two numbers being added. This is a rare property of an operation, which is not enjoyed by division, for instance. Three divided by five is a completely different thing from five divided by three, yet five plus three happens to equal three plus five, and this is not due to any special feature of the numbers three and five; it's true no matter what the numbers are.

What we have here is an example of a universal property—one that is true for all hamsters, not just a special few. One way to indicate this would be to write down a generic formula, such as

$$\square + \triangle = \triangle + \square.$$

The idea is that the box and triangle represent any numbers whatsoever, so the above statement tells us that the sum of two numbers is independent of their order. Operations with this property are said to be *commutative* (from Latin *commutare*, "to exchange"). In any case, this is certainly a property enjoyed by traditional addition: pushing two piles together is a symmetrical activity.

Are there any other universal properties of addition worthy of mention? Actually, there is an important feature of addition that we use almost unconsciously, but it is by no means trivial: we can push several piles together, and it doesn't matter in what order we do it. That is, we can extend the addition activity from two numbers to as many as we like.

Let's say (at least initially) that addition is a binary operation—that is, it assigns a number to a *pair* of numbers only. (We have trained our hamsters to engage in this activity two at a time.) We could then get three of them involved by adding the first two together and then adding the third to that result. Thus, to combine 2, 3, and 7, we would assign to them the entity $(2 + 3) + 7$. Of course, someone else might decide to pair them off in a different way, say $2 + (3 + 7)$. Fortunately, when it comes to addition, these two approaches always lead to the same final result. That is, we have the universal property

$$(\square + \triangle) + \bigstar = \square + (\triangle + \bigstar).$$

This is by no means the case for your random, garden-variety operation. (In fact, it's not even true of subtraction or division.) An operation that satisfies this very convenient property is said to be *associative* (from the Latin *socius*, "companion"). It means that numbers, when being added, get along very well and don't mind being grouped together in whatever order. It also means we can dispense with all the obnoxious parentheses and simply write $2 + 3 + 7$ without ambiguity.

So the addition operation (as an assignment of numbers to number pairs) already distinguishes itself from the herd by possessing these two very nice universal properties. And, as we have seen, multiplication enjoys both of them as well. This means we can add (or multiply) whole bunches of numbers together without having to worry about in what order anything is done. This is an extremely convenient luxury that most operations do not afford us.

> *Define the midpoint operation $\Diamond$ to be*
> *the number halfway between two numbers*
> *(e.g., $3 \Diamond 5 = 4, 9 \Diamond 3 = 6, 4 \Diamond 1 = \frac{5}{2}$).*
> *Is this operation associative?*

Already we see that addition and multiplication set themselves apart from most other operations by their symmetry. In particular, subtraction and division now seem rather unattractive and ungainly by comparison. From a purely aesthetic standpoint, addition is a beautiful princess and subtraction a hideous toad.

In fact, the modern approach would be to dispense with subtraction and division altogether as ultimately redundant and unnecessary operations. This is an amusing and unexpected consequence of the modern mathematical viewpoint, so I'd like to tell you about it.

We saw earlier that it is natural to think of these "lesser" operations as being opposite, or inverse, to their more symmetrical cousins. That is, when we subtract two numbers, say $8 - 5$, we are really asking an addition question: What do you add to 5 to make 8? We could even think of the symbol $8 - 5$ as naming and describing a certain entity: "that which when added to five is eight."

Of course, we already have a name for this creature, namely 3 (as well as many other names such as $1 + 2, 12 \div 4$, and so on). Since 8 is obtained from 3 by the process of adding 5, we can just as well say that 3 comes from 8 by doing the reverse— that is, subtracting 5 can be thought of as unadding 5.

This shows that an inverse process may be less symmetrical and well behaved than the original. Addition is beautifully commutative and associative; unaddition is neither. The same goes for division vis-à-vis multiplication. So that's kind of interesting. It reminds me of many situations in life where undoing something is not as simple or easy as the original doing. Like dropping an egg on the floor, for example. Jigsaw puzzles rely on a fundamental asymmetry: cutting up a photograph into hundreds of small pieces is easy to do, while reassembling them is a time-consuming and laborious process—trying to find that piece with just a little bit of pink on it that forms part of the kitten's nose.

Anyway, we have these operations we want to study and understand, and it becomes convenient (as well as interesting) to have the ability to undo them, but we can't expect the undoings to behave as nicely as the original doings did.

And while we're on the subject, there's an even uglier feature of subtraction that we haven't yet considered: not only does it fail to be symmetrical, it often fails to have any meaning at all. Before we can even start to worry about whether five minus three is the same as three minus five, we have to face the more serious problem of what "three minus five" is supposed to mean in the first place.

From a Piles of Rocks standpoint, such a thing would be nonsensical. How can I take five rocks away when I only have three? I could certainly take three away, leaving none, but then I would still need to take two more. How can I take something from nothing?

In the real world of rocks, of course, I can't. But the nice thing about mathematics is that I get to do whatever I want, whether or not it is physically possible. As long as I can imagine or invent some sort of logically coherent structure, then it immediately becomes part of Mathematical Reality and is every bit as "real" as anything else in mathematics.

The issue here is symmetry—or rather, the lack thereof. With addition, no matter what number I have and no matter what number I want to add to it, their sum is a perfectly valid entity, already extant in the realm of numbers (or hamsters, if you prefer that mental image). With the subtraction operation, however, we have an unpleasant *restriction*: the number we are taking away cannot exceed the number we have. There definitely is already a number that when added to three makes five, but (at the moment) none of our numbers play the role of "the thing that when added to five makes three." This is the sort of unpleasantness that makes a mathematician's nose wrinkle in disgust. (At least mine does, anyway.)

A practical, commonsense sort of person would probably just shrug his shoulders and say that this is simply the way things are. The asymmetry is built in to the nature of subtraction. After all, adding makes things bigger, right? So it stands to reason that you can't take away more than you have. You mathematical dreamers will just have to face reality; you can't always get what you want.

Well, phooey to that. As a matter of fact, I *can* get what I want. The problem appears to be that I am lacking some numbers. So why don't I just make them up? After all, there weren't any fractions until we had the idea to chop up our remainders into pieces. Can't we simply invent some new numbers to play the roles we require? What's to stop us?

This is a major theme in mathematics, actually. We have a structure of some sort—maybe it originated in a real-world context, maybe not—but in any case, we now have a purely abstract imaginary mathematical structure, and we want to investigate its patterns and features. At some point we may discover an unpleasant lack of some sort—the structure feels incomplete in some way, and we wish it were richer, fuller, and thereby more capable of fulfilling its aesthetic purpose: to hold patterned information in the most elegant way possible. This is what is known as an *extension problem*. We want to take our existing structure (in our case, it is the world of numbers and operations) and improve it by adding new entities to it, hopefully without losing any of the nice features and symmetries we already have.

The long and the short of it is that we're going to create some new numbers in order to make subtraction better behaved. I would like to be able to take anything away from anything, and, in particular, I want to be able to take something from nothing. For instance, I would like there to be a number that when added to two makes zero.

One way to do this (at least metaphorically) is to imagine a flock of sheep. Along with the usual, ordinary sort of sheep, I also want to imagine what we might call "antisheep." These are just like regular sheep, except for the curious property that whenever a sheep and an antisheep come into contact, they immediately annihilate each other and disappear. I know there aren't any sheep like that; I'm just pretending. (Actually, our present understanding of the physical universe suggests that such antisheep could very well exist, or be manufactured; it would just be prohibitively expensive.) The point is that now we have something that we can add to a pair of sheep to get none—namely, a pair of antisheep.

We've actually talked about this idea before, with shoes and antishoes. The idea was to trade operations for entities. Just as I can replace the activity of taking off my shoe with putting on my antishoe, we can replace subtraction of sheep with addition of antisheep. For example, if I have five sheep and I wish to remove three of them, one way to do this would be to bring in three antisheep and then let nature take its course.

More abstractly (and much less violently), we can simply bring into existence an entity like "anti-two," which would be defined as "the thing that when added to two is zero." The sheep imagery is in no way necessary; anti-two does not need to be a collection or a number of anything, just an actor with a part to play.

Of course we are talking here about the invention (or discovery, if you prefer) of so-called *negative numbers*—quantities or measurements that are (or can be interpreted as being) less than zero. This has become a fairly commonplace idea, especially in colder climates where the temperature falls well below the (completely arbitrary) zero degree mark on the thermometer. Everyone in North Dakota knows that when it's twenty below it'll need to get twenty degrees warmer just to get to zero.

As ordinary and familiar as all this might seem today, negative quantities were generally held in some suspicion throughout history, regarded as more of a bookkeeping trick than as legitimate numbers in their own right. In fact, the typical Renaissance tradesman never used such things. Income and expenditures were typically recorded separately, using two different colors of ink—black for profit, red for losses (whence the expression "in the red" for being in debt). The black and red numbers could be totaled separately and then compared.

At some point somebody must have noticed that if you perform all the various additions and subtractions that accounting and bookkeeping entail, the final total is independent of the order in which they are done. That is, you don't have to wait until you have enough money in order to spend it (as many of us have since woefully discovered). Going below zero simply means that you owe. Thus, if you like, you can think of negative numbers in *commercial* terms— as measures of debt.

This way of thinking, like the antisheep metaphor, is an example of *reification*—making an abstract idea seem more real or tangible. We have a purely abstract mathematical construct (that is, negative numbers), and in order to feel more comfortable with it, to internalize it in such a way that we can have intuition about it, we come up with these sorts of concrete images and real-world analogies. After all, as intelligent and imaginative as we may be, we're still primates with hands and eyes and experiences, and our minds can use all the help they can get.

So although no such imagery is required to bring these new creatures into existence mathematically, it sure doesn't hurt to have a thermometer or some antisheep in your mind occasionally.

Let's proceed with the plan. For each ordinary "positive" number like 2 or 5, we call into existence a companion "negative" number with the property that when added to it makes zero. (So we have essentially doubled our collection of hamsters.) We could, I suppose, call them black and red numbers and even write them in these colors, but it's a lot simpler (and a lot less expensive) to just use a mark of some kind to distinguish them. The customary notation is the *negative sign* (–), which is admittedly a bit confusing; it looks rather a lot like the subtraction symbol, the minus sign (–). There is good reason for this, as we will see, and it is quite convenient once you get used to it, but it's always a little uncomfortable having the same (or nearly the same) symbol mean two different things, even (and maybe especially) if the ideas they signify are closely related.

The upshot is that we now have numbers such as -2 (pronounced "negative two") and $-\frac{5}{3}$ ("negative five-thirds"), whose only function in life is to annihilate their partner. That is, all that -2 does for a living is to be the thing that when added to 2 makes 0. Thus,

$$2 + (-2) = 0.$$

Notice how the notation makes me want to put -2 in parentheses, to separate the plus (+) symbol from the negative sign. That's pretty lame, actually. The trouble with mathematical notation is that it was often improvised by printers so as to be cost effective; it's easier to use a piece of type you already have than to carve a new one.

In any case, 2 and -2 are like sheep and antisheep. They cancel each other out, so to speak. Notice that this is a symmetrical relationship: not only is -2 acting as an "anti-two," but 2 itself acts to annihilate -2, in effect making it *anti-*anti-two. So we should really think of the negative sign not as a mere label, distinguishing the two kinds of numbers but as a negation operator acting on all numbers, turning each one into its opposite. So -2 means the negation of 2 (that is, negative two as before), whereas $-(-2)$ would mean the negation of negative two, otherwise known as 2.

What is -0?

So now we have both the *subtraction operation*, which involves a pair of numbers, as well as the *negation operator*, acting on a single number. And just to add to the confusion, we use essentially the same symbol for both. On the other hand, there are close connections between the two. For one thing, −2 could be thought of as shorthand for the subtraction 0 − 2. After all, that's what we built it to do—to be the thing that you would get if you took two away from nothing. More generally, subtracting two from anything is the same as adding negative two. It doesn't matter whether we take two sheep away normally or toss in a pair of antisheep instead.

What this means is that we can dispense with subtraction forever. All that subtraction does is to add the opposite, so once everybody has an opposite (which they now do), we can use *adding* to do our subtracting, just as "putting on" is enough for me once I have antishoes at my disposal.

This is good news. Any time we can unload something large and expensive like an operation in exchange for some cheap and abundant mental hamsters, it's worth doing. Now all we have to do is add, and adding is nicer because it's more symmetrical.

In fact, we now get an interesting window into the asymmetry of subtraction: only *one* of the numbers we are adding gets negated. That is, 3 + (−2) is manifestly not symmetric, the 2 being negated and not the 3.

In the modern abstract way of thinking, we do not need a concept of annihilation to say what we mean by negation. The pattern is simply that to each number is associated another number (its negative) with the property that they add up to zero (where zero, recall, is that which when added does nothing).

The charming thing about this viewpoint (which possibly also makes it seem pointless and confusing) is that nowhere do we require that these entities be *quantities* of any kind; that is, they don't have to be numbers, per se. We don't actually care what they are, only how they act. We are content to say that if it behaves the way we need it to then it's good enough. If it quacks like zero, then it *is* zero.

So let us accept these new negative quantities as being every bit as real as their positive counterparts. Mathematically there is no problem, because the whole enterprise is a made-up game in our heads. More practically, we have the various reifications we've discussed, such as antisheep, debt, and thermometers, to ease any occasional queasiness we may have.

The question now is how our new hamsters get along—both with the old ones as well as with each other. The addition situation is fairly clear, especially thinking in terms of sheep: 5 + (−2) = 3, and (−1) + (−1) = −2. Amusingly, there is a new wrinkle to subtraction. If we're saying that subtraction means addition of the opposite, then a consequence would be that subtracting −2 would be the same as adding 2. In other words, removing two antisheep is the same as adding in two more sheep.

An amusing way to see why this makes sense is to imagine adding two sheep and two antisheep to your flock. Since these together add up to nothing, you haven't changed anything. But now if you take the two antisheep away, it is easy to see that the sheep population has just gone up by two.

What is the debt interpretation?

Despite the initial weirdness of addition sometimes making things smaller and subtraction making them bigger, this new system of numbers is actually an excellent improvement and gives us a far prettier, more symmetrical number environment in which to operate. We can now freely add and subtract without any worries or restrictions, every number has an opposite, and all the patterns (e.g., symmetry of addition) remain intact. Moreover, we now have the luxury of relegating subtraction to the trash bin if we wish and can henceforth think of ourselves as adders (not the snake kind).

But what about multiplication and division? When we brought negative numbers into existence, we asked only that they behave in a certain *additive* way—namely, that they combine with their partners to make zero. We neglected to tell them how to multiply. What do we even want $(-2) \times (-3)$ to mean? How am I supposed to make -2 copies of something? I don't know how to do *anything* a negative number of times.

Clearly we are in need of some reification. Ideally, we could use a new copier metaphor that somehow incorporates negation and would help provide intuitive meaning to negative multiplication. There is a nice way to do this, actually, but we have to be a little bit careful. This is again an extension problem—we need to extend the meaning of multiplication to a wider realm. We could, of course, just do it at random. After all, if $(-2) \times (-3)$ has no meaning (and it currently doesn't), then can't we just call it 42 or $-5\frac{1}{2}$ and move on, assigning arbitrary values to all such products at will?

We certainly could do that, but then we run the risk of ruining all of our beautiful patterns and symmetries. The modern approach is to let the patterns themselves dictate our choices. I don't really care what $(-2) \times (-3)$ is chosen to be, so long as it keeps my pretty patterns intact. (The stained-glass window is so beautiful, we'll build the cathedral around it.)

In particular, it would be heartbreaking to lose the symmetry of multiplication. So whatever we choose this number $(-2) \times (-3)$ to be, it had better be the same as $(-3) \times (-2)$ or I'm going to be upset.

That's our first demand—that we keep the symmetry. This actually does force a few decisions. For instance, I know what I want doubling to mean, so that tells me what $2 \times (-1)$ should be: two copies of one antisheep is two antisheep. Thus, $2 \times (-1) = -2$. If we insist on preserving the symmetry (and we do), we are forced to choose $(-1) \times 2$ to mean -2 as well.

Here is our first extension of meaning, and I want to be clear about it. Initially, multiplication meant making copies; 3 $\times$ 5 means "three copies of five." Now we are changing that a bit. We are taking $(-1) \times 2$ to mean not "negative one copies of two" (which is senseless) but rather "whatever it needs to mean to keep the symmetry intact."

Of course, the same argument could be made with 5 in place of 2. Any number of -1's is that many antisheep, so $5 \times (-1) = -5$. In other words, we're simply counting antisheep. (Does that help you stay awake?) Even more generally, if we put an antisheep in the copier and reduce or enlarge it, we get a smaller or larger antisheep. Thus $\frac{2}{3} \times (-1) = -\frac{2}{3}$. A $\frac{2}{3}$-scale antisheep would exactly annihilate a $\frac{2}{3}$-scale sheep, for instance. Preserving the symmetry would force us to choose $(-1) \times 5 = -5$ and $(-1) \times \frac{2}{3} = -\frac{2}{3}$.

So we actually get a fairly nice interpretation: multiplying a number by -1 simply negates it. If we want, we can reinterpret negation as multiplication by -1.

Similarly, multiplication by -2 can also be dealt with in this way. Since five copies of -2 is -10 (I'm thinking antisheep), we would want $(-2) \times 5$ to be the same. Thus, multiplying by -2 acts by doubling *and* negating. This also makes sense if we think of -2 as being $(-1) \times 2$. Then multiplying by -2 is the same as multiplying by both -1 and 2, so we get the effects of both.

It's not hard to extend the copy machine metaphor, now that we see how we want it to behave. Negation is a self-inverse process—to undo it, you just do it again, like flipping over a coin or turning halfway around. Negation is a sort of reflection activity, moving numbers to the other side of zero. So let's model it with actual reflection—that is, in addition to enlarging and reducing, we can allow our copier to make "backward" copies (as I'm sure many real-life copiers are built to do).

Now we have a very simple reification of our extended multiplication operation. Multiplying by a positive number simply scales by that factor as before, but multiplying by a negative number scales and *reverses*.

One consequence of this choice—insisting on symmetry come what may—is that the product of two negative numbers is then forced to be positive. For example, (–2) × (–3) must be equal to 6, since that is what you get when you double and negate –3. Another way to think of it is to imagine feeding the number 1 into the copier, setting it to enlarge by –3 (i.e., to triple and then flip), and then scale again by –2, doubling and flipping again. Clearly this is the same as just enlarging by 6. The point being that all these processes—blowing up, shrinking down, and flipping—can be done in any order, and flipping twice is the same as not flipping at all.

It is not hard to see what the reciprocal of a negative number should be. To undo multiplication by –2, we need only multiply by $-\frac{1}{2}$. The reciprocal of $-\frac{2}{3}$ would have to be $-\frac{3}{2}$. We need the negative sign to kill off the other negative sign. Two wrongs make a right, if you like. (Of course, zero continues to have no reciprocal, for the same reason as before—there's no way to undo such a destructive multiplication.)

If we like, we can then define division by a negative number to be multiplication by its reciprocal, or better yet, dispense with division altogether as redundant and unnecessary.

We now have a completely symmetrical number environment where we are free to add, unadd, multiply, and unmultiply whatever we want, whenever we want. This system, consisting of the positive and negative fractions (which includes, of course, whole numbers like 5, 0, and –3), is known as the *rational number field*. The adjective *rational* here refers to the fact that these numbers are describable as whole number ratios (that is, fractions), not that they are necessarily calm and collected in their reasoning. The rational number field is the proper setting for arithmetic (at least in the perfect mathematical sense) and provides a rich and fertile landscape for investigation and play.

Suppose we wanted there to be two zero entities,
each having the property that when added to any number
leaves it alone. Does anything go wrong?

THE ART OF COUNTING

Now I'd like to show you something really pretty. Let's go back to the simple idea of counting—desiring to know how many of some thing there are. We've talked about the *craft* of arithmetic—the organizing and rearranging of symbolic numerical information for the purpose of comparison. Now it's time to talk about the fine art: the art of counting *beautifully*.

Of course, if the thing you want to count is a random jumble of haphazard information, such as the total of today's receipts or the number of listings in the phone book, then there's really nothing else to do but roll up your sleeves and do some good old-fashioned countin' and arithmatikin'. Not much art there, I'm afraid.

If, however, the thing you are trying to count happens to be very patterned, orderly, and symmetrical in some way, the situation may allow for clever and imaginative ways to count, without the tedium of actually counting. Of course, this may require quite a bit of extra mental labor, but this is all in keeping with the mathematician's general philosophy: *to be willing to think really hard in order to find clever ways to get out of doing any actual work.*

Thus, not only do we spare ourselves the tedium and drudgery of actually counting, we also get the supreme satisfaction of having had an elegant and powerful idea, which will continue to give pleasure long after any utility it may have provided.

So be prepared to see some breathtakingly beautiful ideas. We are counting for counting's sake, and the things we want to count are the *beautiful* things—the things that have a simple and elegant pattern to them.

Let's start with this pretty design.

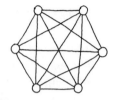

Here is a ring of six dots, with each dot connected by a line to every other dot. You can't get much more patterned and symmetrical than that! The question is, how many lines is that all together?

Of course, we could just count them one by one. If you're reasonably careful you'll get fifteen. But isn't that sort of a wretched existence, being a counter? "Did I forget one? Wait. Did I count that one already? Oh, no. Have I lost track?!"

Not for me, thank you. I simply can't handle the stress. With something this beautiful there has to be a better way. One idea might be to count in an organized fashion that guarantees no lines will be accidentally omitted or counted twice. A natural approach would be to start with one dot, count the lines coming out of it, and then mark them as already counted. Then move on to the next dot, counting the unmarked lines coming out of it, and so forth.

In fact, this is quite pretty. The first dot has five lines, and the next dot will then have four, and so on. That's a nice pattern. We can now see the total number of lines as being

$$5 + 4 + 3 + 2 + 1 = 15.$$

Notice that the final dot adds no lines because they have all been counted already. (So there's really a zero at the end of this sum, if you like.) Clearly, this same idea would work no matter how many dots we had. If there were eleven dots, for instance, the total number of lines would have to be

$$10 + 9 + 8 + 7 + 6 + 5 + 4 + 3 + 2 + 1 = 55.$$

What's great about this idea is that it's not just a count, it's a counting *pattern*. The idea is so simple and robust, it survives the passage to any number of dots whatsoever, no matter how large—or how small, for that matter. For three dots, the pattern would predict 2 + 1 = 3 lines, and that is quite right; a triangle has three sides. Amusingly, the idea keeps working— not just the number pattern; the actual idea itself still makes complete sense as we go to fewer dots. So that's pretty great.

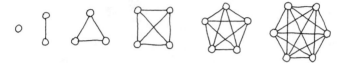

I'd like to show you an even more elegant and amusing way to think about this question. It's a method I like to call "intentional overcounting." The idea is that we're going to make a very simple and easy count, which will be wrong but wrong in a very simple and easily corrected way.

Look at each of the six dots independently. They all have five lines coming out of them. That's six dots, five lines each. Sounds like 6 × 5 to me. But of course this is wrong. We've double-counted some of the lines. As a matter of fact, we've double-counted *all* of the lines (once from each end). This means our initial count of 6 × 5 must be exactly double the correct value, so all we have to do is cut it in half. Thus we get

$$(6 \times 5) \div 2 = 15.$$

I've always loved intentional overcounting. It just feels so deliciously irreverent, somehow. But boy, is it quick! And instead of adding up a long list of consecutive numbers, we just have to multiply a couple of numbers together and divide by two.

Once again, the idea is so general that it applies equally well to any number of dots. If we had eleven dots the thinking would be "eleven dots, ten lines each, makes 110, but I've double-counted, so it's really 55." The same result as before but more gracefully achieved.

Incidentally, by counting the same thing in two different ways, we've uncovered a very pretty arithmetical pattern:

$$1 = (1 \times 2) \div 2$$
$$1 + 2 = (2 \times 3) \div 2$$
$$1 + 2 + 3 = (3 \times 4) \div 2$$
$$1 + 2 + 3 + 4 = (4 \times 5) \div 2$$
$$1 + 2 + 3 + 4 + 5 = (5 \times 6) \div 2$$

The total of the consecutive numbers from 1 up to some number is the same as simply multiplying that number by the next higher number and then taking half.

Another nice way to see this is to imagine making rows of rocks of increasing length to make a triangular design:

This design clearly represents the sum 1 + 2 + 3 + 4 + 5. Now here's the clever part: putting two of these designs together, one upside down, creates a 5 × 6 rectangle!

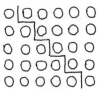

So again we see that the original sum must be the same as half the rectangle, or (5 × 6) ÷ 2.

Notice that we're not really doing a lot of calculation or comparison here. I'm more interested in ways of seeing—

insights into the behavior of numbers in general, not the sizes of anything in particular. For this reason, we won't be doing a lot of arithmetic in the traditional sense. The expression $(5 \times 6) \div 2$, for instance, is actually more valuable and more revealing of the structure of the argument than the calculated value 15. Truth be told, I don't really care how many rocks or dots or lines I've got; I care about the patterns of ideas. I'm happy that I have a way to see that the numbers from 1 to 99 add up to $(99 \times 100) \div 2$, but I don't really care how big that is or what it looks like in Hindu-Arabic decimal form.

What happens when you total only the odd numbers?

For our next example of elegant counting, let's look at something really prosaic and mundane: automobile license plates. How many license plate combinations are there? When I was a kid growing up in California, the license plate format was three letters followed by three numbers, like so:

Since then, due to the completely insane explosion in the number of vehicles, they've had to add another digit at the front, and that has already gotten up to 7 as I write.

Of course, I'm not asking about the number of plates currently out there on the road. That's a totally random collection, contingent upon all sorts of ugly, real-world complexities. I'm interested in the total number of *possible* combinations of three letters and three numbers. This is at least independent of whatever people may be doing with their cars at the moment.

So how are we going to count something massive like this? If you are a masochist, you could start listing the license plates by hand in some way, cross-checking every so often to make sure you haven't accidentally counted anything twice, but then you may as well count the stars in the sky or the hairs on your head. This approach takes no advantage of the symmetry and structure of what we are counting: all possible arrangements of three letters followed by three numbers. While we're at it, we may as well forget about license plates entirely and just view our objects as abstract sequences of symbols. I'm interested in finding a simple and elegant way to count these sequences; I have no interest in what the department of motor vehicles is going to do with them.

As with many mathematical questions, it is often a good idea to start with something smaller and simpler. Let's imagine that our sequences consist of a single letter and a single digit, zero through nine. So we're counting such things as

$$A5, \quad Q9, \quad B7, \quad F0, \quad Z5, \quad T4$$

Making a list of all such combinations is still not something I really want to do. However—and this is really one of the key ideas in the art of counting—there is nothing stopping me from *imagining* such a list. I have no intention of actually writing anything down, mind you, but I can still play around with the mental image of a list.

So the imaginary list technique, as I like to call it, is to see if we can find a way to organize our list (at least in principle) in order to break it into well-regulated and patterned pieces that are easy to count. We can then use arithmetic to assemble the separate counts into a total.

In the present case of single-letter, single-digit license plates, this isn't terribly hard to do. How would an organized, systematic person arrange their imaginary list? A natural way would be to simply list them in alphabetical order—those that begin with A, then the ones that start with B, and so on. In this way, our imaginary list is broken into chapters, or subsections: Chapter A, followed by Chapter B, all the way to Chapter Z.

What makes this organization scheme valuable (as obvious

as it may be) is that it breaks our list into twenty-six chapters of *equal size*. Since every digit is allowed to occur with every letter, each chapter contains exactly ten sequences. (We could even imagine these listed in numerical order, so that Chapter A begins A0, A1, A2, etc.)

This means we can calculate our total without having to list even a single sequence. There are twenty-six chapters, each containing ten entries. So that's $26 \times 10 = 260$ total combinations.

The reason this worked out so nicely is that the sizes of the chapters were all the same, allowing us to use multiplication. If we had broken the list up in a less organized way, into chapters of various sizes, we would then need to add them all together, which might have been something of a chore. So the first principle of making an imaginary list is to break it into equal-size pieces, if possible.

Returning to the original 1960s California setting, let's imagine making a list of all three-letter, three-digit sequences. Working as before, we can still break the list up into chapters depending on the first letter. Then, each of these can be further subdivided according to the second letter. Since each choice of first letter can be followed by all 26 possible second letters, the total number of subsections must be 26×26.

Of course, each of these can be further subdivided by the choices for the third letter (aren't you glad we're not actually doing this?), giving us a total of $26 \times 26 \times 26$ different three-letter categories. In each of these there are exactly one thousand license plates, since the numbers go from 000 to 999. Or, if you prefer, we can think of it as $10 \times 10 \times 10$ by the same subdividing approach we used for the letters.

What this means is that there are exactly

$$26 \times 26 \times 26 \times 10 \times 10 \times 10$$

possible license plate combinations, and our small mental labor has relieved us of a great deal of physical toil. Plus we know we're right—our method guarantees that everybody gets counted exactly once.

Is this enough license plates for twenty million cars?

As another example of this sort of behavior, let's imagine a three-wheel slot machine at a casino (or pretty much anywhere in Nevada). Each wheel has a set of symbols on it, such as cherries, lemons, and lucky sevens, and these show up in a display window like so:

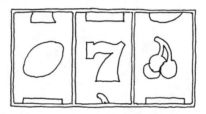

These machines are rather closely related to odometers (sigh) and other mechanical counters, except there's no carry pin. Each wheel is completely independent of the others, and it is this independence that makes things easy to count. The point being that for each possible state of the first wheel, there are exactly the same number of possible combinations for the other two wheels—namely, *all* of them. The first wheel showing a lemon in no way influences or interferes with the other two wheels.

Let's say each wheel has six possible symbols. Then our imaginary list of possible outcomes could be divided into six categories (depending on the first wheel), each of which would have six subsections, and each of these would contain six actual entries, for a total of 6 × 6 × 6 possible combinations.

So here we have a very general and versatile counting principle. If you can find a way to view the things you are counting as being slot-machine possibilities with independent wheels (or slots), then you can calculate the total number of these by simply multiplying together the possibilities for each slot. This is exactly the situation with license plates, where we had three slots with 26 possible entries and three with 10.

Rhode Island plates look like QZ458.
How many plates does that format allow?

While we're on the subject of gambling, it often happens that in addition to enjoying the fun and excitement of a game of chance, people sometimes want to make a rational decision about whether such a game is financially worthwhile—that is, to have some kind of a crude measure as to how likely it is that they will win. Obviously, there can be no guarantees, but on the other hand, some things are more likely to happen than others. One very common way to measure the likelihood of a particular event is to compare the number of ways that event could happen with the total number of possible outcomes.

For example, let's suppose you are playing a slot machine like the one we imagined—three wheels of six symbols each. So that's $6 \times 6 \times 6 = 216$ total possible outcomes. If the only winning combinations were 777 and three cherries, then you would have only 2 chances out of 216. This measurement can be put in the form of a ratio if you like, so that you have a probability of $\frac{2}{216}$, which is about 0.9%, if you prefer that sort of thing. Another way to say it is that the odds against you are 214 to 2, meaning that out of the 216 total possibilities, 214 are against you and only 2 are in your favor. The point is that all these measurements of likelihood come down to counting.

A deck of cards has been shuffled. How likely is it
that the top card is either a spade or a face card?

Now let's look at a more sophisticated counting problem. This one is also connected to my childhood, in particular my interest in stamp collecting, the nations of the world, and especially their various flags. To keep things simple, let's confine ourselves to flags with three vertical stripes, such as those of France and Italy.

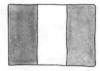

FRANCE   ITALY

Obviously, we're not going to be counting the number of such flags in actual use. There are few things less elegant and mathematical than global politics. (I can't even keep up with all the new countries that are always popping in and out of existence, let alone whatever ghastly flag designs they may have adopted.) Instead, let's suppose we are launching our own new country, and we've narrowed down our choice of colors to these four: red, black, yellow, and green. That is, we've decided on a flag with three vertical stripes, but we haven't yet committed to what color each stripe will be.

If we take the slot-machine view that each stripe is a slot with four possible values, then we would get $4 \times 4 \times 4 = 64$ possible flag designs. But this would include such (presumably unacceptable) colorings as black-black-red and green-green-green, which we did not intend. That is, we want each stripe to be a *different* color, with no repeats. So here we have a restriction on our choices, and that makes things a bit more interesting.

Of course, we can still make an imaginary list and try to break it up into sections in a nice way. In fact, the slot-machine idea even works but with an amusing little twist. We can still subdivide our list of flags into chapters according to the color of the first stripe (say the leftmost one). Now, however, the second stripe can receive only three of the four colors, since we can't repeat. Luckily, though, this number is the same for all choices of the first color. It is true that the set of available colors will depend on what we used for the first stripe (e.g., if it is green, then the second stripe can't be), but the *number* of possible colors is the same, namely three.

So our imaginary list breaks into four chapters, each with three subsections. Finally, each of these subsections contains exactly two flags. This is because after the first two stripes have been colored, there are only two colors left for the third and final stripe. For instance, in the red-black subsection we would find the two designs red-black-yellow and red-black-green.

So, even though we placed a restriction on the colorings, we still have independence of a sort—namely, independence of number. It's as if we have a slot machine where the first wheel has 4 symbols, the second has 3, and the third only 2. The total number of possible flag designs is then $4 \times 3 \times 2 = 24$.

Of course, that is a small enough number that we actually could list them all, and it might even be fun to get out some colored pencils and draw them. I would even recommend doing so if these ideas are at all new to you.

Draw all 24 possible flags.
Which ones do you like best?

Obviously, counting, like any other art form, requires care and patience and years of experience to master. Making imaginary lists and being sensitive to independence is an excellent start.

What if we allow the first and third colors to be the same?

One commonly occurring scenario is when we have a collection of objects—say, books on a bookshelf—and we want to know how many ways there are to put them in order. To keep things simple, let's call the books A, B, C, and D. So now the different orderings will simply be letter sequences, such as DACB and BDCA. How many of these are there?

The first slot has 4 possibilities. Then the second slot has only 3, independent of what the first choice happened to be. That's crucial. Each successive space on the shelf offers one less possibility because the remaining books keep dwindling.

There are two possibilities for the third slot and only one for the fourth. That is, the last book is forced to go at the end, no matter which book it happens to be.

Again, we have number independence, thankfully, and so we get $4 \times 3 \times 2 \times 1 = 24$ possibilities, same as with the striped flags. This is because the colors are essentially acting as books, and the stripes are the places on the bookshelf. The unused color plays the role of the last book.

This happens a lot with counting. Two apparently different counting problems end up being the same. And I don't mean the final tallies are the same; I mean the problems themselves are the same, once we think of them in the right way. That's a lot of what being a mathematician is all about: seeing things in the simplest, most abstract way possible, so you can make connections and understand more deeply. There are few things as powerful and affecting as these kinds of mathematical epiphanies.

We've just discovered a very simple and general counting principle. If you have a bunch of things, say thirteen of them, and you want to put them in order, the number of ways to do this would be

$$13 \times 12 \times 11 \times 10 \times 9 \times 8 \times 7 \times 6 \times 5 \times 4 \times 3 \times 2 \times 1.$$

This sort of product occurs frequently in counting problems, so it's worth abbreviating. The usual way to do this is to write 13!, pronounced "thirteen factorial." The exclamation point is a funny choice, but it has become quite standard. (Again, the printers like to choose from symbols that are already lying around.) The main drawback is that you have to restrain your enthusiasm when writing about numbers; if you write "the answer is 5!" it might be confusing.

Thus, there are $3! = 6$ ways to order three objects, and of course only $2! = 2$ ways to put on your shoes (the usual way and the uncomfortable way with shoes on the wrong feet). Of course $1! = 1$, corresponding to the one and only way to put one thing in order. And if that weren't pedantic enough, we

THE ART OF COUNTING

could even agree to say that $0! = 1$ as well, there being only one way to order an empty bookshelf—do nothing! All in all, it's a very beautiful pattern.

Show there are more than five thousand
ways to order seven things.

What if two of the books are identical
and cannot be distinguished?

Aside from rearranging and permuting objects, another wide class of counting problems involves selecting groups of things from a pool of possibilities. As a simple example, suppose that I need two kids to help me screw in a lightbulb and six kids happen to volunteer. I need to choose two out of the six to be my helpers. How many ways are there to do this?

Obviously, in real life I would just pick two of them (by whatever means) and get on with it. Even then, I could still *wonder* how many choices I actually had. Anyway, we're wondering.

A fantastic way to get out of doing any work is to realize that this is really the same as the dots and lines problem: each dot is a kid, and the lines correspond exactly to the possible pairs. Ha! So it's 15 again.

OK, Mr. Smarty-Pants, suppose then that I need *three* helpers. Groups of three are not so easily captured by dots and lines, so we'll actually need to do some thinking. One clever approach is to intentionally overcount. Let's think of it as having three slots to fill, one for each of my three helpers. There are 6 ways to fill the first slot, then 5 for the second (the first kid is now unavailable), and 4 for the last spot. So that's $6 \times 5 \times 4$ possible choices. Of course, we've overcounted quite drastically, since each group of three has appeared on our list several times. If, for instance, we chose Jane first, then Michael, then Chris, that would get counted separately from choosing Chris first, then Jane, then Michael, even though it results in the exact same team of helpers. Our slot-machine approach to counting has introduced an ordering within the teams that is irrelevant to our problem.

So, how many times did a given team of three get counted? There are exactly 3! = 6 ways to put three people in order, so we've accidentally (on purpose) counted each team exactly six times. That means our total is six times what it should be. So again all we have to do is divide by six. The final count would be

$$(6 \times 5 \times 4) \div 6 = 20.$$

There are twenty ways to choose a group of three out of a group of six. In the counting biz, this is referred to as "six choose three" and is usually abbreviated as $\binom{6}{3}$. Notice this is not a division (there is no crossbar), and the parentheses are a formal part of the symbol, not just a safety precaution.

Thus, $\binom{8}{5}$ represents the number of ways to select five things out of a group of eight, with the tacit understanding that there is no order to anything and no restrictions on which things can be chosen. By the same reasoning as before, we see that this can be calculated as

$$\binom{8}{5} = (8 \times 7 \times 6 \times 5 \times 4) \div 5!$$

The 5! corresponds to the number of possible orderings of five things, which is the factor by which we've overcounted. If you like, we can even evaluate this expression for the purposes of comparison. One simple way to do this is to write out 5! explicitly, then cancel out any common factors in the top and bottom:

$$\frac{8 \times 7 \times 6 \times 5 \times 4}{5 \times 4 \times 3 \times 2 \times 1} = \frac{8 \times 7 \times 6}{3 \times 2 \times 1} = 8 \times 7 = 56.$$

So there are exactly fifty-six ways to choose five things out of a group of eight.

How many ways are there to separate eight people into two groups of four?

Here's a funny one. Suppose I have a box of eight dough-nuts, let's say five chocolate and three coconut. Of course, I can arrange them in the box in any way I like, but I only care about the order of the flavors, not the individual doughnuts themselves. That is, if we let the symbols A and B denote chocolate and coconut, then one box arrange-ment would be

ABAABABA.

Switching two chocolate doughnuts has no effect on this pattern. So I'm wondering, how many sequences of five As and three Bs are there?

Here the idea is to view the *slots* as the objects, not the doughnuts. All I need to do is choose five of the eight spots and put the As in those places. Then the Bs fill in the rest. In this way (without really having to do too much), I can imme-diately deduce the total to be $\binom{8}{5}$.

Why is this the same as $\binom{8}{3}$?

Things get a bit subtler if we add in a third flavor. Let's imagine a box of a dozen doughnuts—five chocolate, three coconut, and four with sprinkles. So we want to count the number of twelve-letter "words" consisting of 5 As, 3 Bs, and 4 Cs. One such word would be

CAABCABACCBA.

See if you can use intentional overcounting to get a handle on this. If we could tell the doughnuts apart, then we would easily get 12! (I mean twelve factorial, not an enthusiastic twelve.)

Can you solve this three-flavor doughnut problem?
(If your reasoning is correct, you should arrive at
a total of 27720 possible arrangements.)

While we're here in the doughnut shop, allow me to show you another of my absolute favorites. This is a perfect example of the kind of creative and imaginative thinking that a good counting problem inspires.

Let's make another box of a dozen doughnuts, only this time we'll allow any number of As, Bs, and Cs, so long as they add up to twelve. I don't really care about the order of the doughnuts in the box, only the amounts of each flavor. So all boxes with 2 As, 3 Bs, and 7 Cs are the same as far as I'm concerned. In fact, we may as well agree to put all the As first, followed by the Bs, and then the Cs.

This means we can rephrase the question: How many ways are there to break 12 into a sum of three numbers? More precisely, we want to count the number of *ordered* triples of numbers that add up to 12. We need to think of 2 + 3 + 7 as different from 7 + 2 + 3 because they correspond to different choices of flavors—the first number indicates the number of As, for instance. This includes the possibility that one or more of the numbers is zero, the sum 6 + 6 + 0 corresponding to a box with 6 As, 6 Bs, and no Cs.

So how are we going to count this collection? We could certainly make an imaginary list and try to organize it in some way. For example, we could break it up by the number of As: first the ones with no As, then the arrangements with a single A, two As, and so on, until the final chapter, consisting of a single arrangement with twelve As and nothing else.

The trouble with this approach is that the chapters are all different sizes. There are tons of arrangements with no As, but only one with twelve. That's going to get messy, especially since the same will be true of the subdivisions within each chapter. Ouch.

So here's the insanely gorgeous idea I want to show you. Let's first imagine that we do something quite practical, which is often done in real-life bakeries and doughnut shops. We will introduce cardboard separators in between the different flavors, so the chocolate doesn't get all over the coconut (not that I would mind, those being pretty much my two favorite substances in the whole wide world). Anyway, a typical box would now look like this:

AAA | BBBBB | CCCC

To be as consistent and symmetrical as possible, let's say that we will *always* insert two separators, even when a flavor is missing. For example, the box with 7 As and 5 Bs (and no Cs) would receive one separator between the As and Bs, and another at the end, like so:

AAAAAAA | BBBBB |

Other amusing examples would be

| | CCCCCCCCCCC
AAAAAA | | CCCCC
| BBBBBBBBBBBB |

No matter what, we always have two separators, and if there are no doughnuts on one side or the other, that's fine.

My favorite thing to do at this point is to imagine the separators not as flimsy pieces of cardboard or waxed paper but as *fake plastic doughnuts*. Then we can think of our boxes as containing fourteen doughnuts, exactly two of which are fake. What's more, *any* positioning of the two fakes is allowed and corresponds to a legal doughnut flavor arrangement.

Now the heavens part and the simple truth is revealed: all we are doing is choosing two things out of fourteen—namely, the two spots where the fake doughnut separators go. So the answer to our question is simply $\binom{14}{2} = 91$. Is that great, or what?

What if there were four flavors to choose from?

I hope I've managed to convey something of the depth, beauty, and subtlety of the art of counting. The best way to appreciate it, of course, is to devise your own counting problems and especially your own clever and witty solutions. I'll leave you with one more gem before I go.

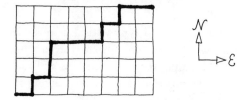

You have arranged to meet a friend at a café
five blocks north and eight blocks east of your house.
Assuming you only walk north and east,
how many different routes are possible?

AFTERWORD

Well, Reader, we've come to the end of our time together. I hope this has all made sense—both the technical details of arithmetic and, more importantly, the larger philosophical and aesthetic perspective. I especially hope that I have managed to get across the idea of viewing your mind as a playground—a place to create beautiful things for your own pleasure and amusement and to marvel at what you've made and at what you have yet to understand. If this book has given you an appreciation for mathematical beauty—on whatever level—then I will be happy indeed.

INDEX

Abacus, 27, 28, 36, 43, 47, 49, 50, 52, 73, 84, 92, 151
Abbreviation, 12, 13
Accuracy, 133, 134
Acre, 2, 79
Adders, 193
Addition, 28, 34, 35, 38, 45, 52, 58, 65, 117-118, 183-187
  symmetry of, 184-187, 192
Aesthetics, 170, 188, 215
Alchemy, 102
Algebraists' creed, 182
Algorithm, 107, 108, 110, 111, 113, 118, 131, 133, 147
  streamlining of, 107, 108, 114, 126, 128
Alien octopodes, 49, 74
Alignment, 60, 65, 68, 70, 83, 107-109, 115, 126, 150
Aliquot part, 154, 156
Alphabetical order, 24, 202
Ampersand (&), 161
Anti-anti-two, 191
Anti-two, 189, 191
Antisheep, 189-195
Antishift, 114-116, 132, 161
Antishoe, 175, 189, 192
Apples and oranges, 115, 164
Approximation, 116, 134, 160, 169, 171, 180-181
Arabic numerals, 51, 61, 75, 155
Arabs, 51, 66, 67, 84
Archimedes, 149
Area, 89
Arithmetic, vii, 16, 20, 24, 91, 97, 104, 117, 151, 197
Arithmetic system, 6, 48
Artisanal approach, 149
Assignment, 184-185
Associative, 185-186
Astronomical, 73, 116
Atom, 153

Austen, Jane, 79, 160
Axle, 140-143

Ba, 15
Babylonians, 21, 49, 77, 78
Baking, 158, 162, 171, 173, 212
Banana People, 15, 21-22, 24, 26, 40, 47, 57, 86, 90, 94, 103, 111, 116
Banana system, 49, 113
Bananas, 16, 17, 21-24
Base-eight. See Octal
Base-four, 112
  See also Quaternary
Base-seven. See Septimal
Base-sixty. See Sexagesimal
Base-ten, 81, 149, 161
  See also Decimal
Base-two, 148
  See also Binary
Bathroom floor, 89
Batting average, 160
Beat, 11, 158
Billion, 73
Billionth, 162
Binary, 114, 148
  See also Base-two
Binary operation, 185
Blowing up, 166-168, 174, 176, 196
Books, 1, 207-209
Borrowing, 69
Box of rocks, 98
Brain, 144
Breakdown strategy, 100, 101, 105, 107, 111
British monetary system, 21, 79, 160
Bushel, 28, 94
Busy box, 19

Café, 214
Calcium, 37
Calculator, 19, 37, 50, 146, 148-150, 151, 169

Calculus stones, 37, 44-45, 49, 52
Capulet, Juliet, 6
Carry gear, 141-142
Carry pin, 141-145, 148, 204
Carrying, 64, 65, 106, 118, 128
Cash register, 147
Cashing in. *See* Exchange
Cats, 6, 37, 44, 49, 52
Cave paintings, 5
Centiliter, 81
Centimeter, 77
Centum, 32, 161
Chalk, 37
Cherries, 204, 205
Chinese abacus, 47
Chinese system, 41, 48, 50
Chocolate, 131, 182, 211, 212
Choose notation, 210
Chutes and Ladders, 9
Clap, 15
Class warfare, 136
Cleopatra, 33
Clockwise, 119, 142, 143
Coconut, 211, 212
Code, 6, 54
Coil of rope, 26, 29, 30
Collection, 5, 183, 211
Column. *See* Place
Combination, 202-204
Common denominator, 166, 168-170, 178
Common sense, 115
Communication, vii, 7
Communist utopia, 120
Commutative, 185-186
Comparative arithmetic, 48
Comparison, vii, 1, 2, 11, 14, 23-24, 29, 56, 90, 93, 113, 117-118, 197
Concubines, 45, 46
Consciousness, 6, 143, 144
Conservation of energy, 125
Consistency, 77, 79, 81, 84, 102, 115, 213
Constellations, 78
Copies, 88, 119, 194
    backward, 195

Copy machine, 167, 172-174, 176, 194-196
Counter, 27, 88, 95, 137, 140, 198, 204
    multidigit, 141
Counterclockwise, 142, 143
Counting, 1, 2, 156, 166, 170, 183, 197, 201, 202, 207-209
Counting coins, 27-28, 37, 95, 121, 123, 157
Counting numbers, 164
Counting pattern, 199
Counting symbols, 43
Cows, 89, 99, 157, 163
Crank, 138, 141, 142, 146
Cratchit, Bob, 80
Creatures, 2, 8, 76, 143, 182
Cross out, 69
Crown, 79
Cup, 2, 152, 158, 173
Curta, 145

Debt, 190, 193
Decaliter, 81
Decameter, 77
Deciliter, 81
Decimal, 94, 113, 116, 127, 159, 165, 178
    See also Base-ten
Decimal currency, 81
Decimal place, 82, 83, 146
    See also Place
Decimal point, 82, 83, 114, 132, 159
Decimeter, 77
Deck of cards, 129, 205
Degrees, 78
Denarius, 79
Denominator, 156-157, 161, 164, 166-167, 176, 178
Density, 133
Desktop, 180, 181
Devangari, 50
Dice, 3, 9
Dickens, Charles, 160
Dictionary, 127
Digit, 52, 74, 103, 109, 115, 128, 140, 142, 148, 149

Digit sequence, 73, 75, 105, 113, 114, 165
Digit wheel, 140-142, 148
Dime, 81, 82, 102, 114, 157, 167
Dinosaurs, 139, 140
Display window, 138, 141, 148, 204
Division, 114, 119, 122, 131, 151, 159, 171, 174-176, 184-186
    asymmetry of, 184-186
Division symbol (÷), 169
Dog, the, 121, 152
Dollar, 81, 82, 102, 114, 157, 167
Dots, 198, 199, 209
Double line, 82
Doubling, 91, 95, 114, 172, 174, 194, 195
Doughnuts, 211-213
Dozen, 12, 49, 88, 99, 157, 166, 212
Drop, 152

Earth, 78, 150
Ecclesiastes, Book of, 119
Eeny, meeny, miney, mo, 120, 121, 122
Eggs, 12, 56, 88, 157, 166, 187
Egyptian system, 25, 26, 33, 48, 94
Egyptians, 25, 26, 41, 48, 91, 95, 102, 121, 153-154
Electronic circuits, 147-149
Eleven, 98
Eleventh-note, 159
Ell, 77
Entertainment, 123, 163
Equals sign (=), 55
Estimation, 115-116, 120, 122, 126-128, 131, 162
Et, 161
Even, 4, 25, 75
Evolution, 7, 139, 140
Exchange, 29, 35, 39, 40, 45, 54, 64, 66-69, 71, 87, 106, 145
Exchange rate, 39, 65, 70
Extension problem, 188, 194

Factorial, 208, 211
Fathom, 77
Fibonacci. See Leonardo of Pisa

Fifteen, 57
Fingers, 7, 9, 12, 52, 67, 78
Finger pattern, 9, 39
Five-barred gate, 11-13, 15, 25, 33, 75, 87
Flags, 205-208
Flipping, 195-196
    See also Negation
Foot, 77, 181
Football players, 6
Fractions, 152, 157, 159, 161, 165, 166, 170, 171, 180, 188
    addition of, 169, 170, 178
    comparison of, 156, 164, 166, 170-171
    division of, 170, 173, 178
    multiplication of, 170, 173, 178
    subtraction of, 169, 170, 178
Fragility, 37, 49
France, 205
Frankfort, Kentucky, 56
French revolution, 76
Furlong, 77

Gallon, 152, 174
Gambling, 205
Generic formula, 184
Goats, 31, 121, 122, 153
Gold, 153
Gram, 82, 83, 133, 134, 159
Grouping, 11, 13, 14, 20, 21, 79, 84, 87, 157
Grouping size, 12-18, 48-49, 67, 77-78, 85, 90, 93, 102, 111, 114, 127, 165-167
Grouping symbols, 33, 36, 42, 43, 49, 52, 73
Guinea, 79

Hamsters, 183-184, 187, 190, 192, 193
Hand People, 15, 17-21, 33, 34, 86, 90
Handful, 2, 11, 12, 14, 15, 18, 34, 120
Hectometer, 77
Heel mark, 26, 29, 30
Hindu mathematicians, 50, 60

Hindu-Arabic system, 8, 51-53, 57, 62, 70, 73, 75-77, 103, 116, 122
Hogshead, 79
Hour, 77, 78
Housing, 138
Hundred, 26, 32, 42, 60, 69, 73, 99, 114
Hundredth, 82, 159, 161, 162

Imaginary list, 202-204, 206-207, 211
Inch, 2, 12, 77
Increment, 64, 137, 141
Independence, 204, 207, 208
India, 50, 51
Infinite loop, 162
Information, vii, 3, 5-7, 11, 20, 22, 50, 56, 70, 148, 176, 188
Information Age, 93
Intentional overcounting, 199, 209, 211
Inverse, 117, 118, 175, 176, 186
Ishango bone, 10
Italy, 205

Japanese, 45, 48, 52, 151
Jelly beans, 5, 119, 120, 121, 125, 128, 131, 152, 170, 171, 174
Jigsaw puzzles, 187
Junk drawer, 149

Khem, 102
Kilogram, 133
Kilometer, 77
Knitting, 30, 64, 72, 151
  See also Symbol knitting

La, 22
Labels, 34
Laboratory, 135, 171
Ladybugs, 37
Language, 6, 7, 12
League, 79
LED (light-emitting diode), 148, 149
Leftovers, 11, 12, 18, 48, 85, 111, 120-121, 122, 133
Lemons, 61, 104, 157, 163, 181, 183, 204

Lentil, 153, 155
Leonardo da Vinci, 145
Leonardo of Pisa, 76
Lexicographic ordering, 24
Liber Abaci, 76
Libra, 79
License plates, 201-205
Life, 143
Liter, 81, 133, 134
Loom137, 151
Losing count, 3
Lotus symbol, 26, 30, 48
Lucky seven, 204, 205
Lumpy, 2, 152, 153

Machine, 136, 137, 139, 144, 149
Maple syrup, 174, 177
Marbles, 1, 5, 13
Marked value, 41, 51, 52, 85, 87, 88, 95, 103
Marked-value system, 22, 25, 29, 32, 33, 35, 91, 97, 102, 111
Mathematical Reality, 163, 182, 187
Mathematics, 7, 97, 99, 119, 150, 163, 178, 181, 182, 187, 188
Meaning, 177
Measurement, 2, 89, 116, 166, 171, 180, 181
Measuring cup, 171
Memorization, 52, 55-57, 92, 93, 112, 113, 136
Memory, 9
Mental arithmetic, 28
Mental column, 72
Mental gymnastics, 50, 87, 92
Meter, 77, 85, 159
Metric system, 76, 82, 84, 86, 159, 161, 171
Midpoint operation, 186
Mile, 77
Milk, 2, 152, 153, 155, 158
Milky Way, 1
Mille, 32
Milligram, 83, 84, 133
Milliliter, 135
Millimeter, 77, 162
Million, 49, 71, 73

Minus sign (−), 55, 191
Minute, 77, 78
Mixed-base system, 21, 79
Model, 180, 181
Modern viewpoint, 117, 118, 132, 182, 186, 192, 194
Money, 1, 6, 22
Muffins, 49, 71, 73
Multicultural, 17
Multiplication, 88, 89, 103–111, 118, 174–176, 195, 203
    symmetry of, 89, 98, 118, 131, 156, 174, 185, 186
Multiplication tables. *See* Times tables
Mutation, 139

Na, 15
Nanometer, 162
Needle, 63–65, 69, 109, 110, 114, 118
Negation, 194–195
Negation operator, 191, 192
    symmetry of, 191
Negative numbers, 189–190, 194–196
    division of, 194, 196
    multiplication of, 194
Negative sign (−), 191, 196
Neuron, 144
Nevada, 204
Nine, 25, 98
North Dakota, 189
Nose, 163, 187, 188
Nothing, 61, 106, 174, 183, 187, 188, 192, 209
Number, 7, 8, 75, 93, 99, 117, 150, 174, 177, 181, 182
Number language, 9, 15, 18, 22, 48, 50, 61
Number permanence, 5
Number sense, 2
Number system, 77, 84
Number words, 85
Numerator, 156, 157, 164, 166, 167–170, 176, 178

Octal, 49, 74, 93, 157
Octillion, 73
Odd, 4, 8, 25, 201

Odds, 205
Odometer, 137, 143, 145, 204
Old Banana, 112
Old World, 79, 150, 151
ON/OFF, 148–149
One, 163, 182, 183
Oneness, 163
Ontological, 67
Operation, 117, 177, 183, 184, 187, 188, 192
Ounce, 79, 153
Overflow, 63

Packed, 23, 38, 45, 58, 91
Part, 153
Pattern, 182, 183, 188, 193, 194, 197, 198, 200, 201, 209
Pattern recognition, 3
Pence, 21, 79, 84, 102
Pencil-and-paper, 52, 87, 92, 113, 131, 147, 149, 151
Pennies, 81, 102, 114, 132, 167
Percent, 161
Percent sign (%), 161
Perception, 3, 11, 14, 16, 25, 32, 47
Perception problem, 10, 14, 20, 25, 33, 41, 48, 49, 73, 164
Perspective, 17, 50, 75, 136, 150, 163, 215
Pharoah, 28, 94, 122
Physical reality, 163, 181
Pickles, 106
Piles of Rocks, 18–20, 27, 50, 92, 151, 157, 187
Pint, 81, 152
Pips, 3, 8
Place, 51, 73, 104, 115
    ones, 46, 51, 59, 64, 83
    hundreds, 51, 59, 70, 71
    tens, 51, 59, 62, 64, 71, 73
    thousands, 59, 65, 71, 72
Place name, 73, 74
Place shifting, 103, 107, 114, 116, 132, 133, 145, 146, 159
    See also Shift
Place value, 36, 37, 41, 51, 87–88, 114, 136, 144

Place-value system, 52, 62, 73, 84, 86, 97, 102, 111, 162
Placeholder, 61
Plato, 181
Plus sign (+), 55, 191
Portability, 6, 44, 49, 52, 103, 149, 181
Positive, 190, 196
Pound, 21, 79
Prime, 8
Printers, 150, 191, 208
Prison, 10, 99
Probability, 205
Product, 90, 98, 105, 109, 118, 146, 174, 208
of sums, 101
Proportion, 171
Pulleys, 125

Quadruple, 92
Quart, 152
Quarter, 152, 157
Quaternary, 112, 113
See also Base-four
Quincunx, 9

Ratchet, 138-140
Ratio, 171, 196, 205
Rational number field, 196
Really, 85
Reciprocal, 174-177
of a negative number, 196
symmetry of, 176
Rectangle, 105-107, 111, 180, 181
of rocks, 88, 98, 100, 101, 108, 118, 177, 200
Reference point, 17, 18
Reflection, 195
Reification, 190, 193, 194
Remainder, 121, 133, 152, 153
Renaming, 166-168
Repetition, 13-14, 32, 41, 48, 49, 87, 97, 111
Representation, 5-8, 18-19, 48, 75, 93, 107, 117-118, 148, 151, 170
Reset, 139, 141
Reverse, 195
Rock of Gibraltar, 44

Rocks, 4, 7, 67, 88, 90, 96, 106, 121, 123, 187, 200
piles of, 8, 54, 87, 117, 183
Roman alphabet, 32, 61
Roman numerals, 33-34, 87
subtractive notation for, 33
Roman system, 32, 33, 35, 41, 50, 52, 76, 87, 102, 103
Romans, 32, 36, 41, 48, 51, 71, 76, 79
Rote, 136, 137, 145
Ruler, 171

Sawyer, Tom, 13
Scale model, 180
Scaling, 166, 168, 171-173, 178, 195
Scaling factor, 168, 172
School, 45, 75, 108, 177
Score, 12
Scribe, 28, 37, 39, 97-99, 154, 161
Scroll, 138-140, 154
Scrooge, Ebenezer, 80
Second, 77, 78
Septimal, 85
Septimal point, 85
Seven-segment display, 148
Sexagesimal, 49
Sharing, 119
Sheep, 6, 12, 13, 189, 191-193
Shift, 103-105, 109, 111-113, 123, 132, 134
See also Place shifting
Shilling, 21, 79, 84, 102
Shrinking down, 167, 174-176, 196
Significant digit, 104
Significant figures, 135
Silver, 79
Silversmith, 94
Silverware, 1
Simplicity, 49
Single-digit numbers, 104, 105, 123, 124, 126, 128
multiples of, 126
products of, 107, 112, 113, 123
sums of, 58, 112
Sixness, 3, 16, 75
Slash (/), 155
Slot machine, 204-206, 209

Smooth, 2, 152, 153
Smythington-Jones, Lady, 80–81
Solar system, 78
Solidus, 79
Soroban, 43–47, 49, 53
Spinning wheel, 151
Sprinkles, 211
Square numbers, 8
Stacking, 25, 32, 33, 39, 48, 49, 50, 87
Stone, 81
Stone patterns, 39
Subatomic particles, 153
Subdivided units, 81, 84, 114, 132, 153, 155, 166
Subgrouping, 39, 41, 48, 49, 50, 71, 87
Subgrouping symbols, 32–34, 36, 41
Subrectangle, 105, 106
Subtraction, 55, 67, 68, 117–118, 183–188
    asymmetry of, 185–187
Sugarplum fairies, 54
Symbols, 7
Symbol knitting, vii, 70
Symmetry, 119, 193–195, 197–198, 202, 213

Tablespoon, 152
Tabula, 36–40, 44, 49, 87, 96, 102
Tally marks, 10, 12, 13, 96, 97
Technology, 139, 149
Ten, 25, 49, 73, 75, 76, 84–85, 90, 103, 111, 112, 161
Tenth, 82, 159, 162
Test tube, 135, 159
Thermometer, 189, 190, 193
Thing-ness, 157
Thinking, 99, 137, 144, 150, 197
Thirtree, 85
Thousand, 26, 42, 58, 63, 73
Thousandth, 83, 159, 160, 162
Three-ness, 2, 8
Thump, 15
Time, 1, 5, 77, 84
Time sense, 3
Times tables, 92, 109, 112–114, 127
Ton, 35
Transistor, 144, 147

Translation, 16, 18, 90, 127
Transmutation, 52, 54, 103, 177
Tree, 16, 85
Tree People, 16, 19, 21, 22, 84, 85, 94
Triangle182, 200
Tribal systems, 48, 86, 87
Trillion, 73
Triode, 147
Triple product, 98
Turkmenistan, 93
Twelve, 17, 78
Two-thirds-ness, 182
Twotree, 85

Unaddition, 118, 119, 186, 196
Units, 2, 77, 79, 82, 152, 158, 171, 179, 180
    imperial, 158, 178
    metric, 77, 86
United States, 159, 178
Universal property, 184, 185
Unmultiplication, 175, 176, 196
Unpacked, 23, 58, 59, 68
Unseven, 175

Vacuum tube, 147
Verb, 118
Vingt, 12
Volume, 98, 135

Waltz, 12
Williams, Ted, 160
Wit, 113, 164, 214
Wolves, 13, 31

Yard, 77
Yard sale, 96
You, 2, 6, 19, 45, 50, 66, 94, 121, 135

Zero, 61–62, 84, 111, 176–177, 183, 196
Zwolf, 12